A PROJECT
MANAGER'S
BOOK OF FORMS

A PROJECT MANAGER'S BOOK OF FORMS

Third Edition

A Companion to the *PMBOK® Guide* – Sixth Edition

Cynthia Snyder Dionisio

Library of Congress Cataloging-in-Publication Data:

Names: Snyder Dionisio, Cynthia, 1962- author.
Title: A project manager's book of forms : a companion to the PMBOK guide –
  sixth edition / Cynthia Snyder.
Description: Third edition. | Hoboken, New Jersey : John Wiley & Sons, Inc.,
  [2017] | Includes index. |
Identifiers: LCCN 2017016425 (print) | LCCN 2017029983 (ebook) | ISBN
  ISBN 9781119393986 (pbk.) | 9781119393993 (pdf) | ISBN 9781119394006 (epub)
  Subjects: LCSH: Project management—Forms.
Classification: LCC HD69.P75 (ebook) | LCC HD69.P75 S689 2017 (print) | DDC
658.4/04—dc23 LC record available at https://lccn.loc.gov/2017016425

SKY10037710_110122

Contents

Acknowledgments   vii

Introduction   ix

New for this Edition   ix

Audience   ix

Organization   x

1   Initiating Forms   1

1.0   Initiating Process Group   /   1

1.1   Project Charter   /   2

1.2   Assumption Log   /   9

1.3   Stakeholder Register   /   12

1.4   Stakeholder Analysis   /   15

2   Planning Forms   17

2.0   Planning Process Group   /   17

2.1   Project Management Plan   /   20

2.2   Change Management Plan   /   25

2.3   Project Roadmap   /   28

2.4   Scope Management Plan   /   30

2.5   Requirements Management Plan   /   33

2.6   Requirements Documentation   /   37

2.7   Requirements Traceability Matrix   /   40

2.8   Project Scope Statement   /   45

2.9   Work Breakdown Structure   /   49

2.10   WBS Dictionary   /   52

2.11   Schedule Management Plan   /   56

2.12   Activity List   /   59

2.13   Activity Attributes   /   62

2.14   Milestone List   /   65

2.15   Network Diagram   /   67

2.16   Duration Estimates   /   70

2.17   Duration Estimating Worksheet   /   73

2.18   Project Schedule   /   78

2.19   Cost Management Plan   /   82

2.20   Cost Estimates   /   85

2.21   Cost Estimating Worksheet   /   88

2.22   Cost Baseline   /   93

2.23   Quality Management Plan   /   95

2.24   Quality Metrics   /   99

2.25   Responsibility Assignment Matrix   /   101

2.26   Resource Management Plan   /   104

2.27   Team Charter   /   109

2.28   Resource Requirements   /   113

2.29   Resource Breakdown Structure   /   116

2.30   Communications Management Plan   /   118

2.31   Risk Management Plan   /   121

2.32   Risk Register   /   128

2.33   Risk Report   /   131

2.34   Probability and Impact Assessment   /   137

2.35   Probability and Impact Matrix   /   142

2.36   Risk Data Sheet   /   144

2.37   Procurement Management Plan   /   147

2.38   Procurement Strategy   /   152

2.39   Source Selection Criteria   /   155

2.40   Stakeholder Engagement
Plan   /   158

3   Executing Forms   163

3.0   Executing Process Group   /   163

3.1   Issue Log   /   165

3.2   Decision Log   /   168

3.3   Change Request   /   170

3.4   Change Log   /   175

3.5   Lessons Learned Register   /   178

3.6   Quality Audit   /   181

3.7   Team Performance
Assessment   /   184

4   Monitoring and Controlling Forms   189

4.0   Monitoring and Controlling Process
Group   /   189

4.1   Team Member Status
Report   /   191

4.2   Project Status Report   /   196

4.3   Variance Analysis   /   202

4.4   Earned Value Analysis   /   206

4.5   Risk Audit   /   209

4.6   Contractor Status Report   /   213

4.7   Procurement Audit   /   218

4.8   Contract Closeout Report   /   222

4.9   Product Acceptance
Form   /   226

5   Closing   229

5.0   Closing Process Group   /   229

5.1   Lessons Learned Summary   /   229

5.2   Project or Phase Closeout   /   235

6   Agile   239

6.1   Product Vision   /   240

6.2   Product Backlog   /   242

6.3   Release Plan   /   244

6.4   Retrospective   /   246

Index   249

Acknowledgments

It was an honor to have chaired the *PMBOK® Guide* – Sixth Edition. My first thanks goes to John Zlockie, PMI Standards Manager, for letting me chair the Sixth Edition. I am so very grateful to Kristin Vitello, Standards Specialist extraordinaire, for being so wonderfully supportive during the entire process. The experience of chairing the Sixth Edition was so rich because of the amazing team of people with whom I worked. I was so honored to have the amazing Dr. David Hillson as my Vice Chair. I am very thankful for his sage guidance and warm friendship. My team was more than I could have hoped for: Mercedes Martinez Sanz, Alejandro Romero Torres, Larkland Brown, Pan Kao, Guy Schleffer, Lynda Bourne, and Gwen Whitman. Frank Parth was with us for a time and his contributions are appreciated. A special place is in our hearts for Michael J. Stratton, who made an incredible difference throughout his life to both the profession and the Sixth Edition. Mike, you are missed.

A huge thank you to my husband Dexter Dionisio for supporting me in the Sixth Edition and in all my writing endeavors. Your support means so much to me.

Thank you to Margaret Cummins. You are such a delight to work with. I always look forward to our times together. I feel so fortunate to work with the fabulous professionals Lauren Olesky, Kalli Schultea, Lauren Freestone, Kerstin Nasdeo, and all the wonderful folks at Wiley. I think the look, feel, and usability of the third edition of the *Book of Forms* will be such an improvement over previous editions. I appreciate the hard work on both the paper and electronic forms.

I appreciate Donn Greenberg, Barbara Walsh, Amy Goretsky, and Roberta Storer for the work you all do to support this book and the other publications we work on together. I am looking forward to the electronic spinoffs for the forms and I appreciate your support with that project!

Thank you to all the students, project managers, and others around the world who have purchased previous editions of this and my other books.

Introduction

A Project Manager's Book of Forms is designed to be a companion to *A Guide to the Project Management Body of Knowledge (PMBOK® Guide)* – Sixth Edition. The purpose is to present the information from the *PMBOK® Guide* – Sixth Edition in a set of forms so that project managers can readily apply the concepts and practices described in the *PMBOK® Guide* to their projects.

The *PMBOK® Guide* identifies that subset of the project management body of knowledge generally recognized as good practice. It does not describe how to apply those practices, nor does it provide a vehicle for transferring that knowledge into practice.

This *Book of Forms* will assist project managers in applying information presented in the *PMBOK® Guide* to project documentation. The *Book of Forms* does not teach project management concepts or describe how to apply project management techniques. Textbooks and classes can fulfill those needs. This book provides an easy way to apply good practices to projects.

NEW FOR THIS EDITION

There are several added features for this edition of the *Book of Forms*. Since one of the defining factors about projects is that they are unique, project managers must tailor the forms and reports to meet the needs of their individual projects. Some projects will require information in addition to what is presented in these forms; some will require less. For each form there is a section that suggests some things to consider for tailoring the form. These forms are presented in paper format and electronic versions to make them easy to adapt to the needs of specific projects. The electronic version is in PDF and .doc format. The .doc format is easy to tailor to suit your needs.

Another section describes the other forms that should be checked for alignment. For example, duration estimates should be aligned with relevant assumptions in the Assumption Log, information in activity attributes, and resource requirements.

Because agile practices are becoming common on projects, even on projects that have not adapted an agile development methodology, we have included a few agile forms. These are not mentioned in the *PMBOK® Guide* – Sixth Edition, but we think they might be helpful.

AUDIENCE

This book is written specifically for project managers to help manage all aspects of the project. Those new to project management can use the forms as a guide in collecting and organizing project information. Experienced project managers can use the forms as a template so that they collect a set of consistent data on all projects. In essence, the forms save reinventing the wheel for each project.

A secondary audience is the manager of project managers or a project management office. Using the information in this book ensures a consistent approach to project documentation. Adopting these forms on an organizational level will enable a repeatable approach to project management.

ORGANIZATION

The forms are organized by Process Group: Initiating, Planning, Executing, Monitoring and Controlling, and Closing. Within those Process Groups, the forms are arranged sequentially as presented in the *PMBOK® Guide* – Sixth Edition. Agile forms are after the closing forms in their own section.

Most form descriptions follow this format:

- A description of each form is presented along with a list of contents. For the planning forms, there is a description of where the information in the form comes from (inputs) and where it goes to (outputs).
- Tailoring. A section that presents information you can consider to help you tailor the forms to fit your needs.
- Alignment. This section presents related forms that you will want to make sure are aligned.
- Description. A table that identifies each of the fields in the form along with a brief explanation.
- A blank copy of the form.

There have been some requests for completed samples of each form. Due to the unique nature of projects and because this book is meant to span all industries and be used by a wide audience, it is not practical to provide examples of completed forms. However, in this current edition we have provided a few samples of filled-out forms so you can see how they would look. These forms are available on the website listed below.

Not all forms will be needed on all projects. Use the forms you need, to the degree that you need them, to assist you in managing your projects.

For electronic copies of the forms, and to see the filled-out examples, go to http://www.wiley.com/go/bookofforms3e.

Initiating Forms

1.0 INITIATING PROCESS GROUP

The purpose of the Initiating Process Group is to authorize a project, provide a high-level definition of the project, and identify stakeholders. There are two processes in the Initiating Process Group:

- Develop project charter
- Identify stakeholders

The intent of the Initiating Process Group is to at least:

- Authorize a project
- Identify project objectives
- Define the initial scope of the project
- Obtain organizational commitment
- Assign a project manager
- Identify project stakeholders

As the first processes in the project, the initiating processes are vital to starting a project effectively. These processes can be revisited throughout the project for validation and elaboration as needed.

The forms used to document initiating information include:

- Project charter
- Assumption log
- Stakeholder register
- Stakeholder analysis

These forms are consistent with the information in the *PMBOK® Guide* – Sixth Edition. Tailor them to meet the needs of your project by editing, combining, or revising them.

1.1 PROJECT CHARTER

The project charter is a document that formally authorizes a project or phase. The project charter defines the reason for the project and assigns a project manager and his or her authority level for the project. The contents of the charter describe the project in high-level terms, such as:

- Project purpose
- High-level project description
- Project boundaries
- Key deliverables
- High-level requirements
- Overall project risk
- Project objectives and related success criteria
- Summary milestone schedule
- Preapproved financial resources
- Key stakeholder list
- Project approval requirements
- Project exit criteria
- Assigned project manager, responsibility, and authority level
- Name and authority of the sponsor or other person(s) authorizing the project charter

The project charter can receive information from:

- Agreements (contracts)
- Statements of work
- Business case
- Benefits management plan

It provides information to:

- Stakeholder register
- Project management plan
- Scope management plan
- Requirements management plan
- Requirements documentation
- Requirements traceability matrix
- Project scope statement
- Schedule management plan
- Cost management plan
- Quality management plan
- Resource management plan
- Communications management plan
- Risk management plan
- Procurement management plan
- Stakeholder engagement plan

The project charter is an output from process 4.1 Develop Project Charter in the *PMBOK® Guide – Sixth Edition*. This document is developed once and is not usually changed unless there is a significant change in the environment, scope, schedule, resources, budget, or stakeholders.

Tailoring Tips

Consider the following tips to help you tailor the project charter to meet your needs:

- Combine the project charter with the project scope statement, especially if your project is small
- If you are doing the project under contract you can use the statement of work as the project charter in some cases

Alignment

The project charter should be aligned and consistent with the following documents:

- Business case
- Project scope statement
- Milestone schedule
- Budget
- Stakeholder register
- Risk register

Description

You can use the element descriptions in Table 1.1 to assist you in developing a project charter.

TABLE 1.1   Elements of a Project Charter

Document Element	Description
Project purpose	The reason the project is being undertaken. May refer to a business case, the organization's strategic plan, external factors, a contract agreement, or any other reason for performing the project.
High-level project description	A summary-level description of the project.
Project boundaries	Limits to the project scope. May include scope exclusions, or other limitations.
Key deliverables	The high-level project and product deliverables. These will be further elaborated in the project scope statement.
High-level requirements	The high-level conditions or capabilities that must be met to satisfy the purpose of the project. Describe the product features and functions that must be present to meet stakeholders' needs and expectations. These will be further elaborated in the requirements documentation.
Overall project risk	An assessment of the overall riskiness of the project. Overall risk can include the underlying political, social, economic, and technological volatility, uncertainty, complexity, and ambiguity. It pertains to the stakeholder exposure to variations in the project outcome.
Project objectives and related success criteria	Project objectives are usually established for at least scope, schedule, and cost. The success criteria identify the metrics or measurements that will be used to measure success.
	There may be additional objectives as well. Some organizations include quality, safety, and stakeholder satisfaction objectives.

(continued)

Table 1.1   Elements of a Project Charter (*continued*)

Document Element	Description
Summary milestone schedule	Significant events in the project. Examples include the completion of key deliverables, the beginning or completion of a project phase, or product acceptance.
Preapproved financial resources	The amount of funding available for the project. May include sources of funding and annual funding limits.
Key stakeholder list	An initial, high-level list of people or groups that have influenced or can influence project success, as well as those who are influenced by its success. This can be further elaborated in the stakeholder register.
Project exit criteria	The performance, metrics, conditions, or other measurements that must be met to conclude the project.
Assigned project manager, responsibility, and authority level	The authority of the project manager with regard to staffing, budget management and variance, technical decisions, and conflict resolution.
	Examples of staffing authority include the power to hire, fire, discipline, accept, or not accept project staff.
	Budget management refers to the authority of the project manager to commit, manage, and control project funds. Variance refers to the variance level that requires escalation.
	Technical decisions describe the authority of the project manager to make technical decisions about deliverables or the project approach.
	Conflict resolution defines the degree to which the project manager can resolve conflict within the team, within the organization, and with external stakeholders.
Name and authority of the sponsor or other person(s) authorizing the project charter	The name, position, and authority of the person who oversees the project manager for the purposes of the project. Common types of authority include the ability to approve changes, determine acceptable variance limits, resolve inter-project conflicts, and champion the project at a senior management level.

PROJECT CHARTER

Project Title: _____

Project Sponsor: _____ Date Prepared: _____

Project Manager: _____ Project Customer: _____

Project Purpose:

```
┌─────────────────────────────────────────────────────────┐
│                                                         │
│                                                         │
│                                                         │
│                                                         │
└─────────────────────────────────────────────────────────┘
```

High-Level Project Description:

```
┌─────────────────────────────────────────────────────────┐
│                                                         │
│                                                         │
│                                                         │
│                                                         │
└─────────────────────────────────────────────────────────┘
```

Project Boundaries:

```
┌─────────────────────────────────────────────────────────┐
│                                                         │
│                                                         │
│                                                         │
└─────────────────────────────────────────────────────────┘
```

Key Deliverables:

```
┌─────────────────────────────────────────────────────────┐
│                                                         │
│                                                         │
│                                                         │
└─────────────────────────────────────────────────────────┘
```

High-Level Requirements:

```
┌─────────────────────────────────────────────────────────┐
│                                                         │
│                                                         │
│                                                         │
└─────────────────────────────────────────────────────────┘
```

Overall Project Risk

```
┌─────────────────────────────────────────────────────────┐
│                                                         │
│                                                         │
│                                                         │
└─────────────────────────────────────────────────────────┘
```

PROJECT CHARTER

Project Objectives	Success Criteria

Scope:

Time:

Cost:

Other:

Summary Milestones	Due Date

PROJECT CHARTER

Preapproved Financial Resources:

| |

Stakeholder(s)	Role

Project Exit Criteria:

| |

Project Manager Authority Level:

Staffing Decisions:

| |

Budget Management and Variance:

| |

PROJECT CHARTER

Technical Decisions:

```
[                                                            ]
```

Conflict Resolution:

```
[                                                            ]
```

Sponsor Authority:

```
[                                                            ]
```

Approvals:

_____   _____

Project Manager Signature   Sponsor or Originator Signature

_____   _____

Project Manager Name   Sponsor or Originator Name

_____   _____

Date   Date

1.2 ASSUMPTION LOG

Assumptions are factors in the planning process that are considered to be true, real, or certain, without proof or demonstration. Constraints are also documented in this log. Constraints are limiting factors that affect the execution of the project. Typical constraints include a predetermined budget or fixed milestones for deliverables. Information in the assumption log includes:

- Identifier
- Category
- Assumption or constraint
- Responsible party
- Due date
- Actions
- Status
- Comments

Assumptions can come from any document in the project. They can also be determined by the project team. Constraints may be documented in the project charter and are determined by the customer, sponsor, or regulatory agencies.

The assumption log provides information to:

- Requirements documentation
- Project scope statement
- Network diagram
- Duration estimates
- Project schedule
- Quality management plan
- Resource estimates
- Risk register
- Stakeholder engagement plan

The assumption log is an output from the process 4.1 Develop Project Charter in the *PMBOK®* *Guide* – Sixth Edition. This log is a dynamic document that is updated throughout the project. Assumptions are progressively elaborated throughout the project and are eventually validated and no longer assumptions.

Tailoring Tips

Consider the following tips to help you tailor the assumption log to meet your needs:

- Combine the assumption log with the issue register and the decision log, to create an AID Log (A = assumption, I = issue, D = decision). You can create them in a spreadsheet with each sheet dedicated to either assumptions, issues, or decisions.
- If you have a very large project you may want to keep the constraints in a separate log from the assumptions.

Alignment

The assumption log should be aligned and consistent with the following documents:

- Project charter
- Issue log
- Risk register

Description

You can use the element descriptions in Table 1.2 to assist you in developing the assumption log.

TABLE 1.2   Elements of an Assumption Log

Document Element	Description
ID	Identifier
Category	The category of the assumption or constraint
Assumption/constraint	A description of the assumption or constraint
Responsible party	The person who is tasked with following up on the assumption to validate if it is true or not
Due date	The date by which the assumption needs to be validated
Actions	Actions that need to be taken to validate assumptions
Status	The status of the assumptions, such as active, transferred, or closed
Comments	Any additional information regarding the assumption or constraint

ASSUMPTION LOG

Project Title: _____ Date Prepared: _____

ID	Category	Assumption/Constraint	Responsible Party	Due Date	Actions	Status	Comments

1.3 STAKEHOLDER REGISTER

The stakeholder register is used to identify those people and organizations impacted by the project and to document relevant information about each stakeholder. Relevant information can include:

- Name
- Position in the organization
- Role in the project
- Contact information
- List of stakeholder's major requirements
- List of stakeholder's expectations
- Classification of each stakeholder

Initially you will not have enough information to complete the stakeholder register. As the project gets underway you will gain additional information and understanding about each stakeholder's requirements, expectations, and classification and the stakeholder register will become more robust.

The stakeholder register receives information from:

- Project charter
- Procurement documents

It is related to:

- Stakeholder analysis matrix

It provides information to:

- Requirements documentation
- Quality management plan
- Communications management plan
- Risk management plan
- Risk register
- Stakeholder engagement plan

The stakeholder register is an output from the process 13.1 Identify Stakeholders in the *PMBOK® Guide* – Sixth Edition. The stakeholder register is a dynamic project document. The stakeholders, their level of influence, requirements, and classification are likely to change throughout the project.

Tailoring Tips

Consider the following tips to help you tailor the stakeholder register to meet your needs:

- Combine the position in the organization with the role on the project, especially if it is a smaller project and everyone knows everyone else's position.
- Combine the stakeholder analysis matrix information with the stakeholder register.
- Eliminate position, role, and contact information for small internal projects.

Alignment

The stakeholder register should be aligned and consistent with the following documents:

- Project charter
- Stakeholder analysis matrix
- Stakeholder engagement plan

Description

You can use the element descriptions in Table 1.3 to assist you in developing the stakeholder register.

TABLE 1.3   Elements of a Stakeholder Register

Document Element	Description
Name	Stakeholder's name. If you don't have a name you can substitute a position or organization until you have more information
Position/Role	The position and/or role the stakeholder holds in the organization. Examples of positions include programmer, human resources analyst, or quality assurance specialist. Roles indicate the function the stakeholder performs on the project team, such as testing lead, Scrum Master, or scheduler
Contact information	How to communicate with the stakeholder, such as their phone number, email address, or physical address
Requirements	High-level needs for the project and/or product
Expectations	Main expectations of the project and/or product
Classification	Some projects may categorize stakeholders as friend, foe, or neutral; others may classify them as high, medium, or low impact

STAKEHOLDER REGISTER

Project Title: _____   Date Prepared: _____

Name	Position/Role	Contact Information	Requirements	Expectations	Classification

1.4 STAKEHOLDER ANALYSIS

Stakeholder analysis is used to classify stakeholders. It can be used to help fill in the stakeholder register. Analyzing stakeholders can also help in planning stakeholder engagement for groups of stakeholders.

The following example is used to assess the relative power (high or low), the relative interest (high or low), and the attitude (friend or foe). There are many other ways to categorize stakeholders. Some examples include:

- Influence/impact
- Power/urgency/legitimacy

Stakeholder analysis receives information from:

- Project charter
- Procurement documents

Stakeholder analysis is a tool used in 13.1 Identify Stakeholders in the *PMBOK® Guide* – Sixth Edition.

Tailoring Tips

Consider the following tips to help you tailor the stakeholder analysis to meet your needs:

- For projects with relatively homogenous stakeholders you can use a 2 × 2 grid that only considers two variables, such as interest and influence.
- For larger projects consider using a 3 × 3 stakeholder cube. Tailor the categories to reflect the importance of various stakeholder variables.

Alignment

The stakeholder analysis should be aligned and consistent with the following documents:

- Stakeholder register
- Stakeholder engagement plan

Description

You can use the element descriptions in Table 1.4 to assist you in developing a stakeholder analysis.

TABLE 1.4   Stakeholder Analysis

Document Element	Description
Name or role	The stakeholder name, organization, or group
Interest	The level of concern the stakeholder has for the project
Influence	The degree to which the stakeholder can drive or influence outcomes for the project
Attitude	The degree to which the stakeholder supports the project

STAKEHOLDER ANALYSIS

Project Title: _____   Date Prepared: _____

Name or Role	Interest	Influence	Attitude

Planning Forms

2.0 PLANNING PROCESS GROUP

The purpose of the Planning Process Group is to elaborate the information from the project charter to create a comprehensive set of plans that will enable the project team to deliver the project objectives. There are 24 processes in the Planning Process Group.

- Develop project management plan
- Plan scope management
- Collect requirements
- Define scope
- Create work breakdown structure (WBS)
- Plan schedule management
- Define activities
- Sequence activities
- Estimate activity durations
- Develop schedule
- Plan cost management
- Estimate costs
- Determine budget
- Plan quality management
- Plan resource management
- Estimate activity resources
- Plan communications management
- Plan risk management
- Identify risks
- Perform qualitative analysis
- Perform quantitative analysis
- Plan risk responses
- Plan procurement management
- Plan stakeholder management

The intent of the Planning Process Group is to at least:

- Elaborate and clarify the project scope
- Develop a realistic schedule

- Develop a realistic budget
- Identify project and product quality processes
- Identify and plan for the needed project resources
- Determine and plan for the communication needs
- Establish risk management practices
- Identify the procurement needs of the project
- Determine how to effectively engage stakeholders
- Combine all the planning information into a project management plan and a set of project documents that are cohesive and integrated

Planning is not a one-time event. It occurs throughout the project. Initial plans will become more detailed as additional information about the project becomes available. Additionally, as changes are approved for the project or product, many of the planning processes will need to be revisited and the documents revised and updated.

Many of the forms in this section provide information needed for other forms. The form description indicates from where information is received and to where it goes.

The forms used to document planning information include:

- Project management plan
- Change management plan
- Project roadmap
- Scope management plan
- Requirements management plan
- Requirements documentation
- Requirements traceability matrix
- Project scope statement
- Assumption log
- Work breakdown structure (WBS)
- Work breakdown structure dictionary
- Schedule management plan
- Activity list
- Activity attributes
- Milestone list
- Network diagram
- Duration estimates
- Duration estimating worksheet
- Project schedule
- Cost management plan
- Cost estimates
- Cost estimating worksheet
- Bottom-up cost estimating worksheet
- Cost baseline
- Quality management plan
- Quality metrics
- Responsibility assignment matrix
- Roles and responsibilities
- Resource management plan
- Team charter
- Resource requirements
- Resource breakdown structure
- Communications management plan

- Risk management plan
- Risk register
- Risk report
- Probability and impact assessment
- Probability and impact matrix
- Risk data sheet
- Procurement management plan
- Procurement strategy
- Source selection criteria
- Procurement strategy
- Stakeholder engagement plan

Some forms in this section are not explicitly described in the *PMBOK® Guide* – Sixth Edition, but they are useful in planning and managing a project. Use the forms here as a starting point for your project. You should tailor the forms to meet the needs of your project by editing, combining, or revising them.

2.1 PROJECT MANAGEMENT PLAN

The project management plan describes how the team will execute, monitor, control, and close the project. While it has some unique information, it is primarily comprised of all the subsidiary management plans and the baselines. The project management plan combines all this information into a cohesive and integrated approach to managing the project. Typical information includes:

- Selected project life cycle
- Development approach for key deliverables
- Variance thresholds
- Baseline management
- Timing and types of reviews

The project management plan contains plans for managing all the Knowledge Areas as well as specific aspects of the project that require special focus. These take the form of subsidiary management plans and can include:

- Change management plan
- Scope management plan
- Schedule management plan
- Requirements management plan
- Cost management plan
- Quality management plan
- Resource management plan
- Communications management plan
- Risk management plan
- Procurement management plan
- Stakeholder engagement plan

The project management plan also contains baselines. Common baselines include:

- Scope baseline
- Schedule baseline
- Cost baseline
- Performance measurement baseline (an integrated baseline that includes scope, schedule, and cost)

In addition, any other relevant, project-specific information that will be used to manage the project is recorded in the project management plan.

The project management plan can receive information from all the subsidiary management plans and baselines. Because it is the foundational document for managing the project it also provides information to all subsidiary plans. In addition, the project management plan provides information to all other integration processes.

The project management plan is an output from the process 4.2 Develop Project Management Plan in the *PMBOK® Guide* – Sixth Edition. This document is developed as the initial project planning is conducted, and then it is not usually changed unless there is a significant change in the charter, environment, or scope of the project.

Tailoring Tips

Consider the following tips to help tailor the project management plan to meet your needs:

- For large and complex projects, each subsidiary management plan will likely be a separate stand-alone plan. In this case you may present your project management plan as a shell with just information on the life cycle, development approach, and key reviews, and then provide a link or reference to the more detailed subsidiary management plans.
- For smaller projects, a project roadmap that summarizes the project phases, major deliverables, milestones, and key reviews may be sufficient.
- You will likely have additional subsidiary management plans that are relevant to the nature of your project, such as a technology management plan, a logistics management plan, a safety management plan, and so forth.

Alignment

The project management plan should be aligned and consistent with the following documents:

- All subsidiary management plans
- Project roadmap
- Milestone list

Description

You can use the element descriptions in Table 2.1 to assist you in developing a project management plan.

TABLE 2.1   Elements of a Project Management Plan

Document Element	Description
Project life cycle	Describe the life cycle that will be used to accomplish the project. This may include the following: • Name of each phase • Key activities for the phase • Key deliverables for the phase • Entry criteria for the phase • Exit criteria for the phase • Key reviews for the phase
Development approaches	Document the specific approach you will take to create key deliverables. Common approaches include predictive approaches, where the scope is known and stable; and adaptive approaches, where the scope is evolving and subject to change. It may also include iterative or incremental development approaches.
Subsidiary management plans	List the subsidiary management plans that are part of the project management plan. This can be in the form of a "table of contents," links to electronic copies of the subsidiary plans, or a list of the other plans that should be considered part of the project management plan, but are separate documents.

(continued)

Table 2.1   Elements of a Project Management Plan (*continued*)

Document Element	Description
Scope variance threshold	Define acceptable scope variances, variances that indicate a warning, and variances that are unacceptable. Scope variance can be indicated by the features and functions that are present in the end product, or the performance metrics that are desired.
Scope baseline management	Describe how the scope baseline will be managed, including responses to acceptable, warning, and unacceptable variances. Define circumstances that would trigger preventive or corrective action and when the change control process would be enacted. Define the difference between a scope revision and a scope change. Generally, a revision does not require the same degree of approval that a change does. For example, changing the color of something is a revision; changing a function is a change.
Schedule variance threshold	Define acceptable schedule variances, variances that indicate a warning, and variances that are unacceptable. Schedule variances may indicate the percent of variance from the baseline or they may include the amount of float used or whether any schedule reserve has been used.
Schedule baseline management	Describe how the schedule baseline will be managed, including responses to acceptable, warning, and unacceptable variances. Define circumstances that would trigger preventive or corrective action and when the change control process would be enacted.
Cost variance threshold	Define acceptable cost variances, variances that indicate a warning, and variances that are unacceptable. Cost variances may indicate the percent of variance from the baseline, such as 0–5 percent, 5–10 percent, and greater than 10 percent.
Cost baseline management	Describe how the cost baseline will be managed, including responses to acceptable, warning, and unacceptable variances. Define circumstances that would trigger preventive or corrective action and when the change control process would be enacted.
Baselines	Attach all project baselines.

PROJECT MANAGEMENT PLAN

Project Title: _____ Date Prepared: _____

Project Life Cycle:

Phase	Key Activities	Key Deliverables

Phase	Reviews	Entry Criteria	Exit Criteria

Development Approaches:

Deliverable	Development Approach

Subsidiary Management Plans:

Name	Comment
Scope	
Time	
Cost	
Quality	
Resource	
Communications	
Risk	
Procurement	
Stakeholder	
Other Plans	

PROJECT MANAGEMENT PLAN

Variance Thresholds:

Scope Variance Threshold	Scope Baseline Management
Schedule Variance Threshold	**Schedule Baseline Management**
Cost Variance Threshold	**Cost Baseline Management**

Baselines:

Attach all project baselines.

2.2 CHANGE MANAGEMENT PLAN

The change management plan is a component of the project management plan. It describes how change will be managed on the project. Typical information includes:

- Structure and membership of a change control board
- Definitions of change
- Change control board
  - Roles
  - Responsibilities
  - Authority
- Change management process
  - Change request submittal
  - Change request tracking
  - Change request review
  - Change request disposition

The change management plan is related to:

- Change log
- Change request form

It provides information to:

- Project management plan

The change management plan is a part of the project management plan and as such it is an output from the process 4.2 Develop Project Management Plan in the *PMBOK® Guide* – Sixth Edition. This document is developed once and is not usually changed.

Tailoring Tips

Consider the following tips to help tailor the change management plan to meet your needs:

- If you have a few product components or project documents that require configuration management, you may be able to combine change management and configuration management into one plan.
- The rigor and structure of your change management plan should reflect the product development approach. For predictive approaches, a rigorous change management approach is appropriate. For adaptive approaches, the change management plan should allow for evolving scope.

Alignment

The change management plan should be aligned and consistent with the following documents:

- Project roadmap or development approach
- Scope management plan
- Requirements management plan

- Schedule management plan
- Cost management plan
- Quality management plan
- Configuration management plan

Description

You can use the descriptions in Table 2.2 to assist you in developing a change management plan.

TABLE 2.2   Elements of a Change Management Plan

Document Element	Description	
Change management approach	Describe the degree of change control and how change control will integrate with other aspects of project management.	
Definitions of change	Schedule Change: Define a schedule change versus a schedule revision. Indicate when a schedule variance needs to go through the change control process to be re-baselined. Budget Change: Define a budget change versus a budget update. Indicate when a budget variance needs to go through the change control process to be re-baselined. Scope Change: Define a scope change versus progressive elaboration. Indicate when a scope variance needs to go through the change control process to be re-basellned. Project Document Change: Define when updates to project management documents or other project documents need to go through the change control process to be re-baselined.	
Change control board	Name	Individual's name
	Role	Position on the change control board
	Responsibility	Responsibilities and activities required of the role
	Authority	Authority level for approving or rejecting changes
Change control process	Change request submittal	Describe the process used to submit change requests, including who receives requests and any special forms, policies, or procedures that need to be used.
	Change request tracking	Describe the process for tracking change requests from submittal to final disposition.
	Change request review	Describe the process used to review change requests, including analysis of impact on project objectives such as schedule, scope, cost, etc.
	Change request outcome	Describe the possible outcomes, such as accept, defer, or reject.

CHANGE MANAGEMENT PLAN

Project Title: _____ **Date Prepared:** _____

Change Management Approach:

Definitions of Change:

Schedule change:
Budget change:
Scope change:
Project document changes:

Change Control Board:

Name	Role	Responsibility	Authority

Change Control Process:

Change request submittal	
Change request tracking	
Change request review	
Change request disposition	

Attach relevant forms used in the change control process.

2.3 PROJECT ROADMAP

The project roadmap is a high-level visual summary of the life cycle phases, key deliverables, management reviews and milestones. Typical information includes:

- Project life cycle phases
- Major deliverables or events in each phase
- Significant milestones
- Timing and types of reviews

The project roadmap can receive information from the project charter and the project management plan. In particular, the key deliverables, project life cycle, management reviews, and scope and schedule baselines.

It provides information to

- Project schedule
- Risk register
- Milestone list

The project roadmap is not listed as an output in the *PMBOK® Guide* – Sixth Edition. If it is developed it would be in conjunction with the project management plan. It is developed once, and then only changed if dates of the key events, milestones, or deliverables change.

Tailoring Tips

Consider the following tips to help tailor the project roadmap to meet your needs:

- For large and complex projects this will likely be a separate stand-alone document.
- For smaller projects the project roadmap may serve as the project management plan.

Alignment

The project roadmap should be aligned and consistent with the following documents:

- Project management plan
- Milestone list

Description

You can use the element descriptions in the following table to assist you in developing a project management plan.

Document Element	Description
Project life cycle phases	The name of each life cycle phase
Major deliverables or events	Key deliverables, phase gates, key approvals, external events, or other significant events in the project
Significant milestones	Milestones in the project
Timing and types of reviews	Management, customer, compliance, or other significant reviews

PROJECT ROADMAP

Approach
Waterfall phases.
Iterative development of content throughout Development, QC Edit, Exposure, and Finalization Phases.

Life Cycle Phases

Timeline

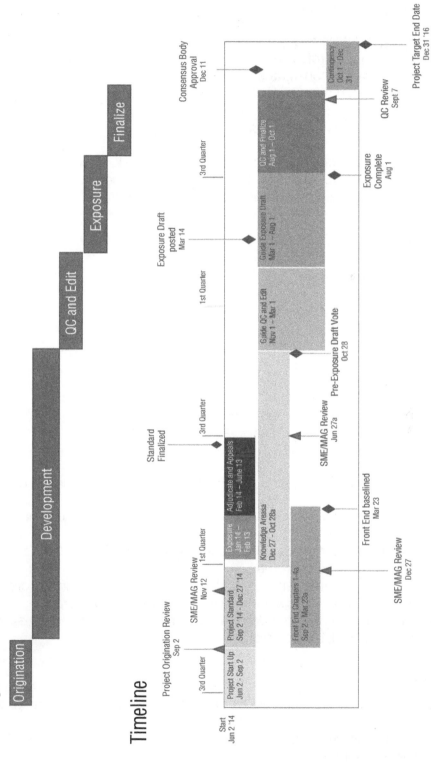

2.4 SCOPE MANAGEMENT PLAN

The scope management plan is part of the project management plan. It specifies how the project scope will be defined, developed, monitored, controlled, and validated. Planning how to manage scope should include at least processes for:

- Developing a detailed scope statement
- Decomposing the project into discrete deliverables using a WBS
- Determining what constitutes a scope change versus a revision and how scope changes will be managed through the formal change control process
- Maintaining the WBS and the scope baseline
- How deliverables will be accepted

In addition, the scope management plan may provide direction on the elements that should be contained in a WBS Dictionary and how the scope and requirements management plans interact.

The scope management plan can receive information from:

- Project charter
- Project management plan

It provides information to:

- Requirements documentation
- Scope statement
- WBS
- WBS dictionary

The scope management plan is an output from the process 5.1 Plan Scope Management in the *PMBOK® Guide* – Sixth Edition. It is developed once and does not usually change.

Tailoring Tips

Consider the following tips to help tailor the scope management plan to meet your needs:

- For smaller projects you can combine the scope management plan with the requirements management plan.
- For larger projects consider a test and evaluation plan that defines how deliverables will be validated and accepted by the customer.
- If your project involves business analysis you may want to incorporate information on how business analysis activities and project management activities will interact.
- If you are using an agile or adaptive development approach you may want to incorporate information on the release and iteration plans.

Alignment

The scope management plan should be aligned and consistent with the following documents:

- Development approach
- Life cycle description
- Change management plan
- Requirements management plan
- Release and iteration plan

Description

You can use the descriptions in Table 2.3 to assist you in developing a scope management plan.

TABLE 2.3   Elements of the Scope Management Plan

Document Element	Description
WBS	Describe the WBS and whether it will be arranged using phases, geography, major deliverables, or some other way. The guidelines for establishing control accounts and work packages can also be documented in this section.
WBS Dictionary	Identify the information that will be documented in the WBS Dictionary and the level of detail required.
Scope baseline maintenance	Identify the types of scope changes that will need to go through the formal change control process and how the scope baseline will be maintained.
Deliverable acceptance	For each deliverable, identify how the deliverable will be validated for customer acceptance, including any tests or documentation needed for sign-off.
Scope and requirements integration	Describe how project and product requirements will be addressed in the scope statement and WBS. Identify the integration points and how requirements and scope validation will occur.
Project management and business analysis integration	Describe how business analysis and project management will integrate as scope is being defined, developed, tested, validated, and turned over to operations.

SCOPE MANAGEMENT PLAN

Project Title: _____ **Date:** _____

Work Breakdown (WBS) Structure

WBS Dictionary

Scope Baseline Maintenance

Deliverable Acceptance

Scope and Requirements Integration

2.5 REQUIREMENTS MANAGEMENT PLAN

The requirements management plan is part of the project management plan. It specifies how requirements activities will be conducted throughout the project. Managing requirements activities includes
at least:

- Planning activities such as:
  - Collecting/eliciting
  - Analyzing
  - Categorizing
  - Prioritizing
  - Documenting
  - Determining metrics
  - Defining the traceability structure
- Managing activities such as:
  - Tracking
  - Reporting
  - Tracing
  - Validating
  - Performing configuration management

The requirements management plan can receive information from:

- Project charter
- Development approach
- Quality management plan

It is related to:

- Scope management plan

It provides information to:

- Requirements documentation
- Requirements traceability matrix
- Quality management plan
- Risk register

The requirements management plan is an output from the process 5.1 Plan Scope Management in the
PMBOK® Guide – Sixth Edition. It is developed once and does not usually change.

Tailoring Tips

Consider the following tips to help tailor the management plan to meet your needs:

- For smaller projects you can combine the requirements management plan with the scope management plan.
- If your project involves business analysis, you may want to incorporate information on how business
  analysis requirements activities and project management requirements activities will interact.
- You can document the test and evaluation strategy in this plan. For larger projects you may want to
  have a separate testing plan.
- If you are using an agile or adaptive development approach you may want to incorporate information
  on how the backlog will be used to manage and track requirements.

Alignment

The requirements management plan should be aligned and consistent with the following documents:

- Development approach
- Change management plan
- Scope management plan
- Release and iteration plan
- Requirements backlog

Description

You can use the descriptions in Table 2.4 to assist you in developing a requirements management plan.

TABLE 2.4   Elements of the Requirements Management Plan

Document Element	Description
Requirements collection	Describe how requirements will be collected or elicited. Consider techniques such as brainstorming, interviewing, observation, etc.
Requirements analysis	Describe how requirements will be analyzed for prioritization, categorization, and impact to the product or project approach.
Requirements categories	Identify categories for requirements such as business, stakeholder, quality, etc.
Requirements documentation	Define how requirements will be documented. The format of a requirements document may range from a simple spreadsheet to more elaborate forms containing detailed descriptions and attachments.
Requirements prioritization	Identify the prioritization approach for requirements. Certain requirements will be non-negotiable, such as those that are regulatory or those that are needed to comply with the organization's policies or infrastructure. Other requirements may be nice to have, but not necessary for functionality.
Requirements metrics	Document the metrics that requirements will be measured against. For example, if the requirement is that the product must be able to support 150 lb, the metric may be that it is designed to support 120 percent (180 lb) and that any design or engineering decisions that cause the product to go below the 120 percent need approval by the customer.
Requirements traceability	Identify the information that will be used to link requirements from their origin to the deliverables that satisfy them.
Requirements tracking	Describe how often and what techniques will be used to track progress on requirements.
Requirements reporting	Describe how reporting on requirements will be conducted and the frequency of such reporting.
Requirements validation	Identify the various methods that will be used to validate requirements such as inspection, audits, demonstration, testing, etc.
Requirements configuration management	Describe the configuration management system that will be used to control requirements, documentation, the change management process, and the authorization levels needed to approve changes.

REQUIREMENTS MANAGEMENT PLAN

Project Title: _____ **Date:** _____

Collection:

Analysis:

Categories:

Documentation:

Prioritization:

REQUIREMENTS MANAGEMENT PLAN

Metrics:

Traceability Structure:

Tracking:

Reporting:

Validation:

Configuration Management:

2.6 REQUIREMENTS DOCUMENTATION

Project success is directly influenced by the discovery and decomposition of stakeholders' needs into requirements and by the care taken in determining, documenting, and managing the requirements of the product, service, or result of the project.

These requirements need to be documented in enough detail to be included in the scope baseline and be measured and validated. Requirements documentation assists the project manager in making tradeoff decisions among requirements and in managing stakeholder expectations. Requirements will be progressively elaborated as more information about the project becomes available.

When documenting requirements, it is useful to group them by category. Some common categories include:

- Business requirements
- Stakeholder requirements
- Solution requirements
- Transition and readiness requirements
- Project requirements
- Quality requirements

Requirements documentation should include at least:

- Identifier
- Requirement
- Stakeholder
- Category
- Priority
- Acceptance criteria
- Test or verification method
- Release or phase

Requirements documentation can receive information from:

- Project charter
- Assumption log
- Stakeholder register
- Scope management plan
- Requirements management plan
- Stakeholder management plan
- Lessons learned register

It provides information to:

- Stakeholder register
- Scope baseline
- Quality management plan
- Resource management plan
- Communications management plan
- Risk register
- Procurement management plan
- Project close out report

Requirements documentation is an output from the process 5.2 Collect Requirements in the *PMBOK® Guide* – Sixth Edition. This process is done once for projects that have a well-defined scope that is not likely to change. For projects that are adaptive, the requirements documentation can evolve and change throughout the project.

Tailoring Tips

Consider the following tips to help tailor the requirements documentation to meet your needs:

- If you are using an agile or adaptive development approach you may want to incorporate information on the release or iteration for each requirement.
- For a project with a lot of requirements you may want to indicate the relationships between requirements.
- You can add information about assumptions or constraints associated with requirements.
- For small and quick adaptive or agile projects, the requirements documentation and backlog can be combined.

Alignment

The requirements documentation should be aligned and consistent with the following documents:

- Requirements management plan
- Benefits management plan
- Quality management plan
- Requirements traceability matrix
- Release plan

Description

You can use the descriptions in Table 2.5 to assist you in developing requirements documentation.

TABLE 2.5   Elements of Requirements Documentation

Document Element	Description
ID	A unique identifier for the requirement
Requirement	The condition or capability that must be met by the project or be present in the product, service, or result to satisfy a need or expectation of a stakeholder
Stakeholder	Stakeholder's name. If you don't have a name you can substitute a position or organization until you have more information
Category	The category of the requirement
Priority	The priority group, for example Level 1, Level 2, etc., or must have, should have, or nice to have
Acceptance criteria	The criteria that must be met for the stakeholder to approve that the requirement has been fulfilled
Test or verification method	The means that will be used to verify that the requirement has been met. This can include inspection, test, demonstration, or analysis
Phase or release	The phase or release in which the requirement will be met

REQUIREMENTS DOCUMENTATION

Project Title: _____ Date Prepared: _____

ID	Requirement	Stakeholder	Category	Priority	Acceptance Criteria	Test or Verification	Phase or Release

2.7 REQUIREMENTS TRACEABILITY MATRIX

A requirements traceability matrix is used to track the various attributes of requirements throughout the project life cycle. It uses information from the requirements documentation and traces how those requirements are addressed through other aspects of the project. The following form shows how requirements would be traced to project objectives, WBS deliverables, the metrics they will be measured to, and how they will be validated.

Another way to use the matrix is to trace the relationship between categories of requirements. For example:

* Business objectives and technical requirements
* Functional requirements and technical requirements
* Requirements and verification method
* Technical requirements and WBS deliverables

An inter-requirements traceability matrix can be used to record this information. A sample form is included after the requirements traceability matrix.

The requirements traceability matrix can receive information from:

* Project charter
* Assumption log
* Stakeholder register
* Scope management plan
* Requirements management plan
* Stakeholder engagement plan
* Lessons learned register

It provides information to:

* Quality management plan
* Procurement statement of work
* Product acceptance
* Change requests

The requirements traceability matrix is an output from the process 5.2 Collect Requirements in the *PMBOK® Guide* – Sixth Edition.

Tailoring Tips

Consider the following tips to help tailor the requirements traceability matrix to meet your needs:

* For complex projects you may need to invest in requirements management software to help manage and track requirements. Using a paper form is usually only helpful for small projects or when tracking requirements at a high level.
* For projects with one or more vendors you may want to add a field indicating which organization is accountable for meeting each requirement.
* Consider an outline format with the business requirement at a parent level and technical requirement and specifications subordinate to the business requirement.

Alignment

The requirements traceability matrix should be aligned and consistent with the following documents:

- Development approach
- Requirements management plan
- Requirements documentation
- Release and iteration plan

Description

You can use the element descriptions in Tables 2.6 and 2.7 to assist you in developing a requirements traceability matrix and an inter-requirements traceability matrix. The matrix shown uses an example of business and technical requirements.

TABLE 2.6   Requirements Traceability Matrix

Document Element	Description
ID	Enter a unique requirement identifier.
Requirement	Document the condition or capability that must be met by the project or be present in the product, service, or result to satisfy a need or expectation of a stakeholder.
Source	The stakeholder that identified the requirement.
Category	Categorize the requirement. Categories can include functional, nonfunctional, maintainability, security, etc.
Priority	Prioritize the requirement category, for example Level 1, Level 2, etc., or must have, should have, or nice to have.
Business objective	List the business objective as identified in the charter or business case that is met by fulfilling the requirement.
Deliverable	Identify the deliverable that is associated with the requirement.
Verification	Describe the metric that is used to measure the satisfaction of the requirement.
Validation	Describe the technique that will be used to validate that the requirement meets the stakeholder needs.

TABLE 2.7   Inter-Requirements Traceability Matrix

Document Element	Description
ID	Enter a unique business requirement identifier.
Business requirement	Document the condition or capability that must be met by the project or be present in the product, service, or result to satisfy the business needs.
Priority	Prioritize the business requirement category, for example Level 1, Level 2, etc., or must have, should have, or nice to have.
Source	Document the stakeholder who identified the business requirement.
ID	Enter a unique technical requirement identifier.
Technical requirement	Document the technical performance that must be met by the deliverable to satisfy a need or expectation of a stakeholder.
Priority	Prioritize the technical requirement category, for example Level 1, Level 2, etc., or must have, should have, or nice to have.
Source	Document the stakeholder who identified the technical requirement.

REQUIREMENTS TRACEABILITY MATRIX

Project Title: _____ Date Prepared: _____

| ID | Requirement Information | | | | | Relationship Traceability | | | |
	Requirement	Source	Priority	Category	Business Objective	Deliverable	Verification	Validation

INTER-REQUIREMENTS TRACEABILITY MATRIX

Project Title: _____ Date Prepared: _____

ID	Business Requirement	Priority	Source	ID	Technical Requirement	Priority	Source

2.8 PROJECT SCOPE STATEMENT

The project scope statement assists in defining and developing the project and product scope. The project scope statement should contain at least this information:

- Project scope description
- Project deliverables
- Product acceptance criteria
- Project exclusions

The project scope statement can receive information from:

- Project charter
- Assumption log
- Scope management plan
- Requirements documentation
- Risk register

It provides information to:

- Work breakdown structure
- Scope baseline

The project scope statement is an output from the process 5.3 Define Scope in the *PMBOK® Guide –* Sixth Edition. It is developed once and is not usually updated unless there is a significant change in scope.

Tailoring Tips

Consider the following tips to help tailor the project scope statement to meet your needs:

- For smaller projects you can combine the project scope statement with the project charter.
- For agile projects you can combine the information with the release and iteration plan.

Alignment

The project scope statement should be aligned and consistent with the following documents:

- Project charter
- Work breakdown structure
- Requirements documentation

Description

You can use the element descriptions in Table 2.8 to assist you in developing a project scope statement.

TABLE 2.8   Elements of a Project Scope Statement

Document Element	Description
Project scope description	Project scope is progressively elaborated from the project description in the project charter and the requirements in the requirements documentation.
Project deliverables	Project deliverables are progressively elaborated from the project description key deliverables in the project charter.
Product acceptance criteria	Acceptance criteria is progressively elaborated from the information in the project charter. Acceptance criteria can be developed for each component of the project.
Project exclusions	Project exclusions clearly define what is out of scope for the product and project.

PROJECT SCOPE STATEMENT

Project Title: _____ Date Prepared: _____

Project Scope Description:

Project Deliverables:

Product Acceptance Criteria:

Project Exclusions:

PROJECT SCOPE STATEMENT

Project Constraints

Project Assumptions

2.9 WORK BREAKDOWN STRUCTURE

The work breakdown structure (WBS) is used to decompose all the work of the project. It begins at the project level and is successively broken down into finer levels of detail. The lowest level, a work package, represents a discrete deliverable that can be decomposed into activities to produce the deliverable.

The WBS should have a method of identifying the hierarchy, such as a numeric structure. The WBS can be shown as a hierarchical chart or as an outline. The approved WBS, its corresponding WBS dictionary, and the project scope statement comprise the scope baseline for the project.

The WBS can receive information from:

- Scope management plan
- Project scope statement
- Requirements documentation

As part of the scope baseline it provides information to:

- Activity list
- Network diagram
- Duration estimates
- Project schedule
- Cost estimates
- Project budget
- Quality management plan
- Resource management plan
- Activity resource requirements
- Risk register
- Procurement management plan
- Accepted deliverables

The work breakdown structure is an output from the process 5.4 Create WBS in the *PMBOK® Guide – Sixth Edition*. The high levels of the WBS are defined at the start of the project. The lower levels may be progressively elaborated as the project continues.

Tailoring Tips

Consider the following tips to help tailor the WBS to meet your needs:

- The needs of the project will determine the way that the WBS is organized. The second level determines the organization of the WBS. Some options for organizing and arranging the WBS include:
  - Geography
  - Major deliverables
  - Life cycle phases
  - Subprojects
- For smaller projects you may use a WBS that is depicted like an organizational chart with deliverables in boxes and subordinate deliverables beneath them.
- For larger projects you will need to arrange the WBS in an outline format and provide a numbering structure.

- At the beginning of the project you may only have two or three levels of the WBS defined. Through the process of progressive elaboration you will continue to decompose the work into more refined deliverables.
- If your organization has accounting codes you may need to align each deliverable to a specific accounting code to track expenditures.

Alignment

The WBS should be aligned and consistent with the following documents:

- Project charter
- Requirements documentation
- Project scope statement
- WBS dictionary
- Activity list

Description

You can use the element descriptions in Table 2.9 to assist you in developing a work breakdown structure.

TABLE 2.9   Elements of a Work Breakdown Structure

Document Element	Description
Control account	The point where scope, schedule, and cost are integrated and used to measure project performance
Work package	The lowest-level deliverable defined in the WBS for estimating and measuring resources, cost, and duration. Each work package rolls up to one and only one control account for reporting purposes.

WORK BREAKDOWN STRUCTURE

Project Title: _____ **Date Prepared:** _____

1. Project
   - 1.1. Major Deliverable
     - 1.1.1. Control Account
       - 1.1.1.1. Work Package
       - 1.1.1.2. Work Package
       - 1.1.1.3. Work Package
     - 1.1.2. Work Package
   - 1.2. Control Account
     - 1.2.1. Work Package
     - 1.2.2. Work Package
   - 1.3. Major Deliverable
     - 1.3.1. Control Account
       - 1.3.1.1. Work Package
       - 1.3.1.2. Work Package
       - 1.3.1.3. Work Package
     - 1.3.2. Control Account
       - 1.3.2.1. Work Package
       - 1.3.2.2. Work Package

2.10 WBS DICTIONARY

The WBS dictionary supports the work breakdown structure (WBS) by providing detail about the control accounts and work packages it contains. The dictionary can provide detailed information about each work package or summary information at the control account level. The approved WBS, its corresponding WBS dictionary, and the project scope statement comprise the scope baseline for the project. Information in the WBS dictionary can include:

- Code of account identifier
- Description of work
- Assumptions and constraints
- Responsible organization or person
- Schedule milestones
- Associated schedule activities
- Resources required
- Cost estimates
- Quality requirements
- Acceptance criteria
- Technical information or references
- Agreement (contract) information

The WBS dictionary is progressively elaborated as the planning processes progress. Once the WBS is developed, the statement of work for a particular work package may be defined, but the necessary activities, cost estimates, and resource requirements may not be known. Thus, the inputs for the WBS dictionary are more detailed than for the WBS.

Use the information from your project to tailor the form to best meet your needs.

The WBS dictionary can receive information from:

- Requirements documentation
- Project scope statement
- Assumption log
- Activity list
- Milestone list
- Activity resource requirements
- Cost estimates
- Quality metrics
- Contracts

As part of the scope baseline, the WBS dictionary provides information to:

- Activity list
- Network diagram
- Duration estimates
- Project schedule
- Cost estimates
- Project budget
- Quality management plan
- Resource management plan
- Activity resource requirements
- Risk register

- Procurement management plan
- Accepted deliverables

The WBS dictionary is an output from the process 5.4 Create WBS in the *PMBOK® Guide* – Sixth Edition. It is progressively elaborated throughout the project.

Tailoring Tips

Consider the following tips to help tailor the WBS dictionary to meet your needs:

- For smaller projects you may not need a WBS dictionary.
- For projects that do use a WBS dictionary you can tailor the information to be as detailed or as high level as you need. You may just want to list a description of work, the cost estimate, key delivery dates, and assigned resources.
- For projects that have deliverables outsourced you can consider the WBS dictionary as a mini-statement of work for the outsourced deliverables.
- Projects that use WBS dictionaries can reference other documents and the relevant sections for technical, quality, or agreement information.

Alignment

The WBS dictionary should be aligned and consistent with the following documents:

- Project charter
- Requirements documentation
- Project scope statement
- WBS
- Activity list

Description

You can use the element descriptions in Table 2.10 to assist you in developing a WBS dictionary.

TABLE 2.10   Elements of a WBS Dictionary

Document Element	Description
Work package name	Enter a brief description of the work package deliverable from the WBS.
Code of account	Enter the code of account from the WBS.
Milestones	List any milestones associated with the work package.
Due dates	List the due dates for milestones associated with the work package.
ID	Enter a unique activity identifier—usually an extension of the WBS code of accounts.
Activity	Describe the activity from the activity list or the schedule.
Team resource	Identify the resources, usually from the resource requirements.
Labor hours	Enter the total effort required.
Labor rate	Enter the labor rate, usually from cost estimates.
Labor total	Multiply the effort hours times the labor rate.
Material units	Enter the amount of material required, usually from the resource requirements.
Material cost	Enter the material cost, usually from cost estimates.
Material total	Multiply the material units times the material cost.
Total cost	Sum the labor, materials, and any other costs associated with the work package.
Quality requirements	Document any quality requirements or metrics associated with the work package.
Acceptance criteria	Describe the acceptance criteria for the deliverable, usually from the scope statement.
Technical information	Describe or reference any technical requirements or documentation needed to complete the work package.
Agreement information	Reference any contracts or other agreements that impact the work package.

WBS DICTIONARY

Project Title: _____   Date Prepared: _____

Work Package Name:	Code of Accounts:

Description of Work:	Assumptions and Constraints:

Milestones:

Due Dates:

1.

2.

3.

ID	Activity	Resource	Labor			Material			Total Cost
			Hours	Rate	Total	Units	Cost	Total	

Quality Requirements:

Acceptance Criteria:

Technical Information:

Agreement Information:

2.11 SCHEDULE MANAGEMENT PLAN

The schedule management plan is part of the project management plan. It specifies how the project schedule will be developed, monitored, and controlled. Planning how to manage the schedule can include at least:

- Scheduling methodology
- Scheduling tool
- Level of accuracy for duration estimates
- Units of measure
- Variance thresholds
- Schedule reporting information and format
- Organizational procedure links
- Schedule updates

The schedule management plan can receive information from:

- Project charter
- Project management plan

It provides information to:

- Activity list
- Activity attributes
- Network diagram
- Activity duration estimates
- Project schedule
- Schedule baseline
- Risk register

The schedule management plan is an output from the process 6.1 Plan Schedule Management in the *PMBOK® Guide* – Sixth Edition. It is developed once and does not usually change.

Tailoring Tips

Consider the following tips to help tailor the schedule management plan to meet your needs:

- Add information on the level of detail and timing for WBS decomposition based on rolling wave planning.
- For projects that use agile, add information on the time box periods for releases, waves, and iterations.
- For projects that use earned value management, include information on rules for establishing percent complete and the EVM measurement techniques (fixed formula, percent complete, level or effort, etc.).

Alignment

The schedule management plan should be aligned and consistent with the following documents:

- Project charter
- Cost management plan

Description

You can use the descriptions in Table 2.11 to assist you in developing a schedule management plan.

TABLE 2.11   Elements of the Schedule Management Plan

Document Element	Description
Schedule methodology	Identify the scheduling methodology that will be used for the project, whether it is critical path, agile, or some other methodology.
Scheduling tool(s)	Identify the scheduling tool(s) that will be used for the project. Tools can include scheduling software, reporting software, earned value software, etc.
Level of accuracy	Describe the level of accuracy needed for estimates. The level of accuracy may evolve over time as more information is known (progressive elaboration). If there are guidelines for rolling wave planning and the level of refinement that will be used for duration and effort estimates, indicate the levels of accuracy required as time progresses.
Units of measure	Indicate whether duration estimates will be in days, weeks, months, or some other unit of measure.
Variance thresholds	Indicate the measures that determine whether an activity, work package, or the project as a whole is on time, requires preventive action, or is late and requires corrective action.
Schedule reporting and format	Document the schedule information required for status and progress reporting. If a specific reporting format will be used, attach a copy or refer to the specific form or template.
Organizational procedure links	The schedule outline should follow the numbering structure of the WBS. It may also need to follow the organization's code of accounts or other accounting and reporting structures.
Schedule updates	Document the process for updating the schedule, including update frequency, permissions, and version control. Indicate the guidelines for maintaining baseline integrity and for re-baselining if necessary.

SCHEDULE MANAGEMENT PLAN

Project Title: _____ Date: _____

Schedule Methodology:

```



```

Scheduling Tools:

```



```

Level of Accuracy:	Units of Measure:	Variance Thresholds:

Schedule Reporting and Format:

```



```

Organizational Procedure Links:

```



```

Schedule Updates:

```



```

2.12 ACTIVITY LIST

The activity list defines all the activities necessary to complete the project work. It also describes the activities in sufficient detail so that the person performing the work understands the requirements necessary to complete it correctly. The activity list contains:

- Activity identifier
- Activity name
- Description of work

The activity list can receive information from:

- Schedule management plan
- Scope baseline (particularly the deliverables from the WBS)

It provides information to:

- Network diagram
- Activity duration estimates
- Gantt chart or other schedule presentation
- Activity resource requirements

The activity list is an output from process 6.2 Define Activities in the *PMBOK® Guide* – Sixth Edition. This process is performed throughout the planning of the project. For adaptive approaches, or projects that are progressively elaborated, this process is done throughout the project.

Tailoring Tips

Consider the following tips to help tailor the activity list to meet your needs:

- For some projects you will enter your activities into the schedule tool rather than keep a separate activity list.
- Not all projects require a column to describe the work.
- For projects that use an adaptive development approach your activity list will evolve as the requirements are added or changed in the product backlog.
- For projects that use an adaptive development approach you may want to add a column that indicates the planned release or iteration for each activity.

Alignment

The activity list should be aligned and consistent with the following documents:

- Milestone list
- Activity attributes
- WBS
- WBS dictionary
- Product backlog
- Iteration release plan

Description

You can use the element descriptions in Table 2.12 to assist you in developing an activity list.

TABLE 2.12   Elements of an Activity List

Document Element	Description
ID	Unique identifier
Activity name	A brief statement that summarizes the activity. Activities usually start with a verb and are only a few words.
Description of work	If needed use this field to provide more detail to the activity description, such as a process or method to accomplish the work.

ACTIVITY LIST

Project Title: _____ Date Prepared: _____

ID	Activity	Description of Work

2.13 ACTIVITY ATTRIBUTES

Activity attributes are the details about the activity. Sometimes the information is entered directly into the schedule software. Other times the information is collected in a form that can be used later to assist in building the schedule model. Activity attributes can include:

- Activity identifier or code
- Activity name
- Activity description
- Predecessor and successor activities
- Logical relationships
- Leads and lags
- Imposed dates
- Constraints
- Assumptions
- Resource requirements and skill levels
- Location of performance
- Type of effort

The activity attributes are progressively elaborated as the planning processes progress. Once the activity list is complete, the description of work for a particular activity may be defined but the necessary attributes, such as logical relationships and resource requirements, may not be known. Thus, the inputs for the activity attributes are more detailed than for the activity list and are added to as new information becomes available.

The activity attributes can receive information from:

- Schedule management plan
- Activity list
- Network diagram
- Scope baseline
- Assumption log
- Activity resource requirements

It provides information to:

- Network diagram
- Duration estimates
- Project schedule
- Resource requirements

Activity attributes are an output from process 6.2 Define Activities in the *PMBOK® Guide* – Sixth Edition. They are developed throughout the planning of the project. For adaptive approaches, or projects that are progressively elaborated, they are updated throughout the project.

Tailoring Tips

Consider the following tips to help tailor the activity attributes to meet your needs:

- For some projects you will enter your activities into the schedule tool rather than keep a separate set of activity attributes.
- Only include the fields you feel necessary to effectively manage your project.
- Constraints and assumptions can be recorded in the assumption log rather than this form.

- For projects that use an adaptive development approach your activity list will evolve as the requirements are added or changed in the product backlog.
- For projects that use an adaptive development approach you may want to add a field that indicates the planned release or iteration for each activity.

Alignment

The activity attributes should be aligned and consistent with the following documents:

- Assumption log
- WBS
- WBS dictionary
- Milestone list
- Activity list
- Network diagram
- Duration estimates
- Schedule
- Cost estimates
- Resource requirements

Description

You can use the descriptions in Table 2.13 to assist you in developing the activity attributes.

TABLE 2.13   **Elements of Activity Attributes**

Document Element	Description
ID	Unique identifier
Activity name	Document a brief statement that summarizes the activity. The activity name starts with a verb and is usually only a few words.
Description of work	A description of the activity in enough detail that the person(s) performing the work understands what is required to complete it.
Predecessor and successor activities	Identify any predecessor activities that must occur before the activity. Identify any successor activities that can't occur until after the activity.
Logical relationships	Describe the nature of the relationship between predecessor or successor activities, such as start-to-start, finish-to-start, or finish-to-finish.
Leads and lags	Any required delays between activities (lag) or accelerations (lead) that apply to the logical relationships
Imposed dates	Note any required dates for start, completion, reviews, or accomplishments.
Constraints	Document any limitations associated with the activity, such as finish-no-later-than dates, approaches to work, resources, etc.
Assumptions	Document any assumptions associated with the activity, such as availability of resources, skill sets, or other assumptions that impact the activity.
Team resources and skill levels	Document the number and roles of people needed to complete the work along with the skill level, such as junior, senior, etc.
Required physical resources	Document the materials, supplies or equipment needed to complete the activity.
Location of performance	If the work is to be completed somewhere other than at the performing site, indicate the location.
Type of effort	Indicate if the work is a fixed duration, fixed effort, level of effort, apportioned effort, or other type of work.

ACTIVITY ATTRIBUTES

Project Title: _____ Date Prepared: _____

ID:	Activity:				

Description of Work:

Predecessors	Relationship	Lead or Lag	Successor	Relationship	Lead or Lag

Number and Type of Team Resources Required:	Skill Requirements:	Required Resources:

Type of Effort:

Location of Performance:

Imposed Dates or Other Constraints:

Assumptions:

2.14 MILESTONE LIST

The milestone list defines all the project milestones and describes the nature of each one. It may categorize the milestone as optional or mandatory, internal or external, interim or final, or in any other way that supports the needs of the project.

The milestone list can receive information from:

- Project charter
- Schedule management plan
- Scope baseline

It provides information to:

- Network diagram
- Duration estimates
- Gantt chart or other schedule presentation
- Change requests
- Close out report

The milestone list is an output from process 6.2 Define Activities in the *PMBOK® Guide* – Sixth Edition. This is developed once and is not usually changed unless there is a significant scope change.

Tailoring Tips

Consider the following tip to help tailor the milestone list to meet your needs:

- For some projects you will enter your milestones directly into the schedule tool rather than keep a separate list of milestones.

Alignment

The milestone list should be aligned and consistent with the following documents:

- Project charter
- Activity list
- Network diagram
- Duration estimates
- Schedule

Description

You can use the descriptions in Table 2.14 to assist you in developing the milestone list.

TABLE 2.14   Elements of a Milestone List

Document Element	Description
Milestone name	Milestone name that uniquely defines the milestone
Milestone description	A description of the milestone in enough detail to understand what is needed to determine the milestone is complete
Type	A description of the type of milestone, such as • Internal or external • Interim or final • Mandatory or optional

MILESTONE LIST

Project Title: _____ Date Prepared: _____

Milestone	Milestone Description	Type

2.15 NETWORK DIAGRAM

The network diagram is a visual display of the relationship between schedule elements. The purpose is visually depict the types of relationships between components. The components are shown nodes that connected by lines with arrows that indicate the nature of the relationship. Relationships can be one of types:

1. Finish-to-start (FS). This is the most common type of relationship. The predecessor elem be complete before the successor element can begin.
2. Start-to-start (SS). In this relationship, the predecessor element must begin befo element begins.
3. Finish-to-finish (FF). In this relationship, the predecessor element must b suc-cessor element can be complete.
4. Start-to-finish (SF). This is the least common type of relationship nt must begin before the predecessor element can be complete.

In addition to the types of relationships, the network d now modifications to the relation-ships, such as leads or lags:

- A *lag* is a directed delay betwee ents. In a finish-to-start relationship with a three-day lag, the successor activity would not until three days after the predecessor was complete. This would be shown as FS+3d. Lag is oat.
- A *lead* is an acceleration ween elements. In a finish-to-start relationship with a three-day lead, the successor activity would begin three days before the predecessor was complete. This would be shown as FS–3d.
- Leads and lags can be applied to any type of relationship.

The network diagram can receive information from:

- Assumption log
- Schedule management plan
- Activity list
- Activity attributes
- Milestone list
- Scope baseline

It provides information to:

- Project schedule

The network diagram is an output from the process 6.3 Sequence Activities in the *PMBOK® Guide – Sixth Edition*. This is developed once and is not usually changed unless there is a significant scope change.

Tailoring Tips

Consider the following tips to help tailor the network diagram to meet your needs:

- At the start of the project you may want to use sticky notes to create a network diagram to get the idea of the flow of deliverables.

- For some projects you will enter the type of relationship directly into the schedule tool rather than draw it out. Most scheduling software has the option to see the schedule as a network diagram.
- The network diagram can be produced at the activity level, the deliverable level, or the milestone level.

Alignment

The network diagram should be aligned and consistent with the following documents:

- Project schedule
- Project roadmap
- Milestone list

NETWORK DIAGRAM

Project Title: _____   **Date Prepared:** _____

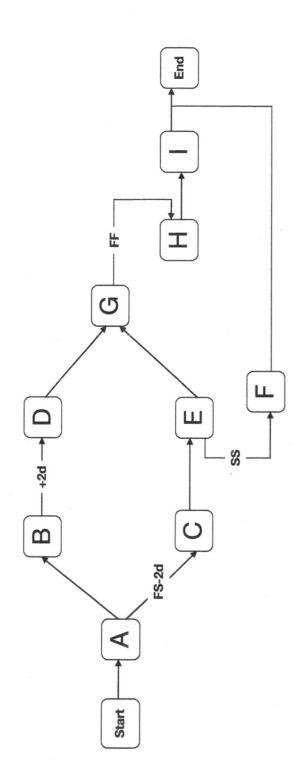

In this network diagram:

There is a two-day lead between the completion of A and beginning of C.

There is a two-day lag between the completion of B and beginning of D.

There is a start-to-start relationship between E and F.

There is a finish-to-finish relationship between G and H.

All other relationships are finish-to-start.

Page 1 of 1

2.16 DURATION ESTIMATES

Duration estimates provide information on the amount of time it will take to complete project work. They can be determined by developing an estimate for each activity, work package, or control account, using expert judgment or a quantitative method, such as:

- Parametric estimating
- Analogous estimating
- Three-point estimating

For those activity durations driven by human resources, as opposed to material or equipment, the duration estimates will generally convert the estimate of effort hours into days or weeks. To convert effort hours into days, take the total number of hours and divide by 8. To convert to weeks, take the total number of hours and divide by 40.

Duration estimates include at least:

- ID
- Activity description
- Effort hours

Duration estimates can receive information from:

- Assumption log
- Scope baseline
- Schedule management plan
- Activity list
- Activity attributes
- Milestone list
- Resource requirements
- Resource breakdown structure
- Resource calendars
- Project team assignments
- Resource breakdown structure
- Risk register
- Lessons learned register

It provides information to:

- Schedule baseline
- Project schedule
- Risk register

Duration estimates are an output from process 6.4 Estimate Activity Durations in the *PMBOK® Guide –* Sixth Edition. They are developed throughout the project as schedule and activity details are refined.

Tailoring Tips

Consider the following tips to help tailor the duration estimates to meet your needs:

- Duration estimates may include contingency reserve to account for risks related to uncertainty in the duration estimates, ambiguity in the scope, or resource availability.

- Duration estimates at the level of accuracy that suits your project needs. Rolling wave planning is often used for duration estimating; as more information is known about the project activities, duration estimates are refined and updated.
- For projects that are using an agile development approach, set time boxes are used rather than duration estimates. Additionally, different estimating methods are used.

Alignment

The duration estimates should be aligned and consistent with the following documents:

- Assumption log
- Activity attributes
- Resource requirements

Description

You can use the descriptions in Table 2.15 to assist you in developing the duration estimates.

TABLE 2.15   Elements of Duration Estimates

Document Element	Description
ID	Unique identifier
Activity description	A description of the work that needs to be done
Effort hours	The amount of labor it will take to accomplish the work; usually shown in hours but may be shown in days
Duration estimates	The length of time it will take to accomplish the work; usually shown in days may be shown in weeks or months

DURATION ESTIMATES

Project Title: _____ Date Prepared: _____

ID	Activity Description	Effort Hours	Duration Estimate

2.17 DURATION ESTIMATING WORKSHEET

A duration estimating worksheet can help to develop duration estimates when quantitative methods are used. Quantitative methods include:

- Parametric estimates
- Analogous estimates
- Three-point estimates

Parametric estimates are derived by determining the effort hours needed to complete the work. The effort hours are then calculated by:

- Dividing the estimated hours by resource quantity (i.e., number of people assigned to the task)
- Dividing the estimated hours by the percent of time the resource(s) are available (i.e., 100 percent of the time, 75 percent of the time, or 50 percent of the time)
- Multiplying the estimated hours by a performance factor. Experts in a field generally complete work faster than people with an average skill level or novices. Therefore, a factor to account for the productivity is developed.

Duration estimates can be made even more accurate by considering that most people are productive on project work only about 75 percent of the time.

Analogous estimates are derived by comparing current work to previous similar work. The size of the previous work and the duration is compared to the expected size of the current work compared to the previous work. Then the ratio of the size of the current work is multiplied by the previous duration to determine an estimate. Various factors, such as complexity, can be factored in to make the estimate more accurate. This type of estimate is generally used to get a high-level estimate when detailed information is not available.

A three-point estimate can be used to account for uncertainty in the duration estimate. Stakeholders provide estimates for optimistic, most likely, and pessimistic scenarios. These estimates are put into an equation to determine an expected duration. The needs of the project determine the appropriate equation, but a common equation is the beta distribution:

$$\text{Estimated duration} = \frac{\text{Optimistic duration} + 4\,\text{most likely duration} + \text{pessimistic duration}}{6}$$

In formulas, duration is often represented by "t" for "time."
The duration estimating worksheet can receive information from:

- Assumption log
- Scope baseline
- Activity list
- Activity attributes
- Resource requirements
- Resource calendars
- Project team assignments
- Risk register
- Lessons learned register

It provides information to:

- Duration estimates

Duration estimates are an output from process 6.4 Estimate Activity Duration in the *PMBOK® Guide –* Sixth Edition. They are developed throughout the project as schedule and activity details are refined.

Description

You can use the element descriptions in Table 2.16 to assist you in estimating durations with the worksheet.

TABLE 2.16   Elements of an Activity Duration Estimating Worksheet

Document Element	Description
ID	Unique identifier
Parametric estimates	
Effort hours	Enter amount of labor it will take to accomplish the work. Usually shown in hours, but may also be shown in days. Example: 150 hours
Resource quantity	Document the number of resources available. Example: 2 people
Percent available	Enter amount of time the resources are available. Usually shown as the percent of time available per day or per week. Example: 75 percent of the time
Performance factor	Estimate a performance factor if appropriate. Generally effort hours are estimated based on the amount of effort it would take the average resource to complete the work. This can be modified if you have a highly skilled resource or someone who has very little experience. The more skilled the resource, the lower the performance factor. For example, an average resource would have a 1.0 performance factor. A highly skilled resource could get the work done faster, so you multiply the effort hours times a performance factor of .8. A less skilled resource will take longer to get the work done, so you would multiply the effort hours times 1.2. Example: A skilled worker with a performance factor of .8
Duration estimate	Divide the effort hours by the resource quantity times the percent available times the performance factor to determine the length of time it will take to accomplish the work. The equation is: $$\text{Effort}/(\text{quantity} \times \text{percent available} \times \text{performance factor}) = \text{duration}$$ Example: $150/(2 \times .75 \times .8) = 125$ hours
Analogous estimates	
Previous activity	Enter a description of the previous activity. Example: Build a 160 square foot deck.
Previous duration	Document the duration of the previous activity. Example: 10 days
Current activity	Describe how the current activity is different. Example: Build a 200 square foot deck.
Multiplier	Divide the current activity by the previous activity to get a multiplier. Example: 200/160 = 1.25
Duration estimate	Multiply the duration for the previous activity by the multiplier to calculate the duration estimate for the current activity. Example: 10 days x 1.25 = 12.5 days
Three-point estimate (Beta distribution)	
Optimistic duration	Determine an optimistic duration estimate. Optimistic estimates assume everything will go well and there won't be any delays in material and that all resources are available and will perform as expected. Example: 20 days

(continued)

TABLE 2.16 Elements of an Activity Duration Estimating Worksheet (*continued*)

Document Element	Description
Most likely duration	Determine a most likely duration estimate. Most likely estimates assume that there will be some delays but nothing out of the ordinary. Example: 25 days
Pessimistic duration	Determine a pessimistic duration estimate. Pessimistic estimates assume there are significant risks that will materialize and cause delays. Example: 36 days
Weighting equation	Weight the three estimates and divide. The most common method of weighting is the beta distribution: $tE = (tO + 4tM + tP)/6$ Example: $(20 + 4(25) + 36)/6$
Expected duration	Enter the expected duration based on the beta distribution calculation. Example: 26 days

DURATION ESTIMATING WORKSHEET

Project Title: _____ Date Prepared: _____

Parametric Estimates

ID	Effort Hours	Resource Quantity	% Available	Performance Factor	Duration Estimate

Analogous Estimates

ID	Previous Activity	Previous Duration	Current Activity	Multiplier	Duration Estimate

Three-Point Estimates

ID	Optimistic Duration	Most Likely Duration	Pessimistic Duration	Weighting Equation	Expected Duration Estimate

2.18 PROJECT SCHEDULE

The project schedule combines the information from the activity list, network diagram, resource requirements, duration estimates, and any other relevant information to determine the start and finish dates for project activities. A common way of showing a schedule is via Gantt chart showing the dependencies between activities. The sample Gantt chart is for designing, building, and installing kitchen cabinets. It shows the:

- WBS identifier
- Activity name
- Start dates
- Finish dates
- Resource name (next to the bar)

The project schedule can receive information from:

- Assumption log
- Schedule management plan
- Activity list
- Activity attributes
- Milestone list
- Network diagram
- Duration estimates
- Project team assignments
- Resource calendars
- Project scope statement
- Risk register
- Lessons learned register

It provides information to:

- Cost estimates
- Project budget
- Resource management plan
- Risk register
- Stakeholder engagement plan

The project schedule is an output from the process 6.5 Develop Schedule in the *PMBOK® Guide –* Sixth Edition. It is updated and elaborated throughout the projects.

Tailoring Tips

Consider the following tips to help tailor the project schedule to meet your needs:

- For smaller projects you may not require scheduling software; you can use a spreadsheet or some other means to show the schedule.
- For projects that use scheduling software, the needs of your project will determine the fields you use and the information you enter and track.
- At the beginning of the project you may only enter the information from the WBS into the schedule. As you further decompose the work you will enter activities, durations, resources, and other information. The level of detail depends on the needs of the project.

- You may choose to only distribute a milestone chart that shows only the dates of the important events or key deliverables. The sample milestone chart is for constructing a house. It shows the activity milestones as well as their dependencies. Showing dependencies on a milestone chart is optional.

Alignment

The project schedule should be aligned and consistent with the following documents:

- Project charter
- Assumption log
- Schedule management plan
- Project roadmap
- Scope baseline
- Activity list
- Network diagram
- Duration estimates
- Project team assignments
- Project calendars

Sample Gantt Chart

ID	WBS	Task Name	Start	Finish
1	**1**	**Kitchen Cabinets**	**Aug 4**	**Oct 2**
2	**1.1**	**Preparation**	**Aug 4**	**Aug 20**
3	1.1.1	Design kitchen layout	Aug 4	Aug 8
4	1.1.2	Design cabinet layout	Aug 6	Aug 12
5	1.1.3	Select materials	Aug 13	Aug 15
6	1.1.4	Purchase materials	Aug 18	Aug 20
7	1.1.5	Preparation complete	Aug 20	Aug 20
8	**1.2**	**Construction**	**Aug 21**	**Sep 26**
9	1.2.1	Build cabinet framing	Aug 21	Sep 10
10	1.2.2	Stain and finish cabinet framing	Sep 11	Sep 12
11	1.2.3	Make cabinet doors	Sep 11	Sep 24
12	1.2.4	Stain and finish doors	Sep 25	Sep 26
13	1.2.5	Make doors	Sep 11	Sep 17
14	1.2.6	Stain and finish drawers	Sep 18	Sep 18
15	1.2.7	Make shelving	Sep 11	Sep 16
16	1.2.8	Stain and finish shelving	Sep 17	Sep 17
17	1.2.9	Construction complete	Sep 26	Sep 26
18	**1.3**	**Installation**	**Sep 29**	**Oct 2**
19	1.3.1	Install cabinet framing	Sep 29	Oct 1
20	1.3.2	Install cabinets	Oct 2	Oct 2
21	1.3.3	Install drawers	Oct 2	Oct 2
22	**1.4**	**Sign off**	Oct 2	Oct 2

Sample Milestone Chart

ID		Task Name	Finish
1		Vendors selected	Mar 3
2		Financing obtained	Mar 3
3		Plans complete	Apr 11
4		Permits obtained	May 2
5		Paving complete	May 2
6		Foundation complete	May 14
7		House framed	Jun 13
8		Roof set	Jun 20
9		Power established	Jun 20
10		Power complete	Jul 11
11		Plumbing complete	Aug 22
12		HVAC complete	Aug 22
13		Finish work complete	Sep 26
14		Garden site prepared	Oct 10
15		City sign-off	O
16		Punch list closed	

2.19 COST MANAGEMENT PLAN

The cost management plan is a part of the project management plan. It specifies how the project costs will be estimated, structured, monitored, and controlled. The cost management plan can include the following information:

- Level of accuracy for cost estimates
- Units of measure
- Variance thresholds
- Rules for performance measurement
- Cost reporting information and format
- Process for estimating costs
- Process for developing a time-phased budget
- Process for monitoring and controlling costs

In addition, the cost management plan may include information on the level of authority associated with cost and budget allocation and commitment, funding limitations, and options and guidelines on how and when costs get recorded for the project.

The cost management plan can receive information from the:

- Project charter
- Schedule management plan
- Risk management plan

It provides information to:

- Activity cost estimates
- Cost baseline
- Risk register

The cost management plan is an output from the process 7.1 Plan Cost Management in the *PMBOK®* *Guide* – Sixth Edition. It is developed once and does not usually change.

Tailoring Tips

Consider the following tips to help tailor the cost management plan to meet your needs:

- On smaller projects, often the project manager does not manage the budget. In those cases you would not need this form.
- Units of measure for each type of resource may be indicated in the cost management plan or the resource management plan.
- For projects that use earned value management, include information on rules for establishing percent complete, the EVM measurement techniques (fixed formula, percent complete, level or effort, etc.). For those that don't, delete this field.

Alignment

The cost management plan should be aligned and consistent with the following documents:

- Project charter
- Schedule management plan

Description

You can use the descriptions in Table 2.17 to assist you in developing a cost managemer

TABLE 2.17   Elements of a Cost Management Plan

Document Element	Description
Units of measure	Indicate how each type of resource will be measured. For exar ... iay be measured in staff hours, days, or weeks. Physical resourc ... ed in gallons, meters, tons, or whatever is appropriate for the mat ... es are based on a lump sum cost each time they are used.
Level of precision	Indicate whether cost estimates will be rounded to hundr ... ome other measurement.
Level of accuracy	Describe the level of accuracy needed for estimates. T' ... ay evolve over time as more information is known (progr ... ere are guidelines for rolling wave planning and the level of ... ed for cost estimates, indicate the levels of accuracy req'
Organizational procedure links	Cost estimating and reporting should follow the ... e WBS. It may also need to follow the organization's cod ... ounting and reporting structures.
Control thresholds	Indicate the measures that determine whet! ... je, or the project as a whole is on budget, requires ... budget and requires corrective action. Usually indica* ... om the baseline.
Rules of performance measurement	Identify the level in the WBS where pro ... be measured. For projects that use earned value manag ... ts will be reported at the work package or control accou ... surement method that will be used, such as weighted ... percent complete, etc. Document the equations that will b ... s to complete (ETC) and estimates at completion (EAC)
Cost reporting information and format	Document the cost information re ... gress reporting. If a specific reporting format will be used ... o the specific form or template. Indicate the reporting frequ
Additional details	Describe variables associated ... oices, such as make or buy, buy or lease, borrowing funds ... unding, etc.

COST MANAGEMENT PLAN

Project Title: _____ Date Prepared: _____

Units of Measure:	Level of Precision:	Level of Accuracy:

Organizational Procedure Links:

Control Thresholds:

Rules of Performance Measurement:

Cost Reporting and Format:

Additional Details:

2.20 COST ESTIMATES

Cost estimates provide information on the cost of resources necessary to complete project work, including labor, equipment, supplies, services, facilities, and material. Estimates can be determined by developing an approximation for each work package using expert judgment or by using a quantitative method such as:

- Parametric estimates
- Analogous estimates
- Three-point estimates

Cost estimates should include at least:

- ID
- Labor costs
- Physical resource costs
- Reserve
- Estimate
- Basis of estimates
- Method
- Assumptions
- Range
- Confidence level

Cost estimates can receive information from:

- Cost management plan
- Scope baseline
- Project schedule
- Quality management plan
- Resource requirements
- Risk register
- Lessons learned register

They provide information to:

- Cost baseline
- Resource requirements
- Risk register

Cost estimates are an output from the process 7.2 Estimate Costs in the *PMBOK® Guide* – Sixth Edition. Cost estimates are developed and then refined periodically as needed.

Tailoring Tips

Consider the following tips to help tailor the cost estimates to meet your needs:

- Cost estimates may include contingency reserve to account for risks related to uncertainty in the Cost estimates or ambiguity in the scope or resource availability.
- If considerations for the cost of quality, cost of financing, or indirect costs were included, add that information to your cost estimate.

- Estimate costs at the level of accuracy and precision that suits your project needs. Rolling wave planning is often used for cost estimating; as more information is known about the scope and resources, cost estimates are refined and updated.
- If using vendors, indicate the estimated cost and indicate the type of contract being used to account for possible fees and awards.

Alignment

The cost estimates should be aligned and consistent with the following documents:

- Assumption log
- Activity attributes
- Project schedule
- Resource requirements
- Project team assignments

Description

You can use the descriptions in Table 2.18 to assist you in developing the cost estimates.

TABLE 2.18   Elements of an Activity Cost Estimate

Document Element	Description
ID	Unique identifier, such as the WBS ID or activity ID
Resource	The resource (person, equipment, material) needed for the WBS deliverable
Labor costs	The costs associated with team or outsourced resources
Physical costs	Costs associated with material, equipment, supplies, or other physical resources
Reserve	Document contingency reserve amounts, if any
Estimate	The sum of the cost of labor, physical resources, and reserve costs
Basis of estimates	Information such as cost per pound, duration of the work, square feet, etc.
Method	The method used to estimate the cost, such as analogous, parametric, etc.
Assumptions/constraints	Assumptions used to estimate the cost, such as the length of time the resource will be needed
Range	The range of estimate
Confidence level	The degree of confidence in the estimate

ACTIVITY COST ESTIMATES

Project Title: _____ **Date Prepared:** _____

WBS ID	Resource	Labor Costs	Physical Costs	Reserve	Estimate	Method	Assumptions/ Constraints	Basis of Estimates	Range	Confidence Level

2.21 COST ESTIMATING WORKSHEET

A cost estimating worksheet can help to develop cost estimates when quantitative methods or a bottom-up estimate are developed. Quantitative methods include:

- Parametric estimates
- Analogous estimates
- Three-point estimates

Parametric estimates are derived by determining the cost variable that will be used and the cost per unit. The number of units is multiplied by the cost per unit to derive a cost estimate.

Analogous estimates are derived by comparing current work to previous similar work. The size of the previous work and the cost are compared to the expected size of the current work. Then the ratio of the size of the current work compared to the previous work is multiplied by the previous cost to determine an estimate. Various factors, such as complexity and price increases, can be factored in to make the estimate more accurate. This type of estimate is generally used to get a high-level estimate when detailed information is not available.

A three-point estimate can be used to account for uncertainty in the cost estimate. Stakeholders provide estimates for optimistic, most likely, and pessimistic scenarios. These estimates are put into an equation to determine an expected cost. The needs of the project determine the appropriate equation, though a common equation is a beta distribution:

$$\text{Estimated cost} = \frac{\text{Optimistic Cost} + 4\,(\text{Most Likely Cost}) + \text{Pessimistic Cost}}{6}$$

Bottom-up estimates are detailed estimates done at the work package level. Detailed information on the work package, such as technical requirements, engineering drawings, labor duration, and other direct and indirect costs are used to determine the most accurate estimate possible.

The cost estimating worksheet can receive information from:

- Cost management plan
- Scope baseline
- Project schedule
- Quality management plan
- Resource requirements
- Risk register
- Lessons learned register

Cost estimating worksheets are process 7.2 Estimate Costs in the *PMBOK® Guide* – Sixth Edition. Cost estimating worksheets are developed and then refined as periodically as needed.

Description

You can use the element descriptions in Table 2.19 to assist you in developing a cost estimating worksheet and the element descriptions in Table 2.20 to assist you in developing a bottom-up cost estimating worksheet.

TABLE 2.19   Elements of a Cost Estimating Worksheet

Document Element	Description
ID	Unique identifier, such as the WBS ID or activity ID
Parametric estimates	
Cost variable	Enter the cost estimating driver, such as hours, square feet, gallons, or some other quantifiable measure. Example: Square feet
Cost per unit	Record the cost per unit. Example: $9.50
Number of units	Enter the number of units. Example: 36
Cost estimate	Multiply the number of units times the cost per unit to calculate the estimate. Example: $9.50 x 36 = $342
Analogous estimates	
Previous activity	Enter a description of the previous activity. Example: Build a 160 square foot deck.
Previous cost	Document the cost of the previous activity. Example: $5,000
Current activity	Describe how the current activity is different. Example: Build a 200 square foot deck.
Multiplier	Divide the current activity by the previous activity to get a multiplier. Example: 200/160 = 1.25
Cost Estimate	Multiply the cost for the previous activity by the multiplier to calculate the Cost Estimate for the current activity. Example: $5,000 x 1.25 = $6,250
Three-point estimate (Beta distribution)	
Optimistic cost	Determine an optimistic cost estimate. Optimistic estimates assume all costs were identified and there won't be any cost increases in material, labor, or other cost drivers. Example: $4,000
Most likely cost	Determine a most likely cost estimate. Most likely estimates assume that there will be some cost fluctuations but nothing out of the ordinary. Example: $5,000
Pessimistic cost	Determine a pessimistic cost estimate. Pessimistic estimates assume there are significant risks that will materialize and cause cost overruns. Example: $7,500
Weighting equation	Weight the three estimates and divide. The most common method of weighting is the beta distribution, where c = cost: $$c_E = (c_O + c_{4M} + c_P)/6$$ Example: $\left(4{,}000 + 4(5{,}000)\right)/6$
Expected cost	Enter the expected cost based on the beta distribution. Example: $5,250

You can use the element descriptions in the following table to assist you in developing a bottom-up cost estimating worksheet.

TABLE 2.20   Elements of a Bottom-Up Cost Estimating Worksheet

Document Element	Description
ID	Unique identifier, such as the WBS ID or activity ID
Labor hours	Enter the estimated effort hours.
Labor rate	Enter the hourly or daily rate.
Total labor	Multiply the labor hours times the labor rate.
Material	Enter quotes for material, either from vendors or multiply the amount of material times the cost per unit.
Supplies	Enter quotes for supplies, either from vendors or multiply the amount of supplies times the cost per unit.
Equipment	Enter quotes to rent, lease, or purchase equipment.
Travel	Enter quotes for travel.
Other direct costs	Enter any other direct costs and document the type of cost.
Indirect costs	Enter any indirect costs, such as overhead.
Reserve	Enter any contingency reserve cost for the work package.
Estimate	Sum the labor, materials, supplies, equipment, travel, other direct costs, indirect costs, and any contingency reserve.

COST ESTIMATING WORKSHEET

Project Title: _____ Date Prepared: _____

Parametric Estimates

ID	Cost Variable	Cost per Unit	Number of Units	Cost Estimate

Analogous Estimates

ID	Previous Activity	Previous Cost	Current Activity	Multiplier	Cost Estimate

Three-Point Estimates

ID	Optimistic Cost	Most Likely Cost	Pessimistic Cost	Weighting Equation	Expected Cost Estimate

BOTTOM-UP COST ESTIMATING WORKSHEET

Project Title: _____   Date Prepared: _____

ID	Labor Hours	Labor Rate	Total Labor	Material	Supplies	Equipment	Travel	Other Direct Costs	Indirect Costs	Reserve	Estimate

2.22 COST BASELINE

The cost baseline is a time-phased budget that is used to measure, monitor, and control cost performance for the project. It is developed by summing the costs of the project by the time period and developing a cumulative cost curve that can be used to track actual performance, planned performance, and the funds spent.

A project may have multiple cost baselines; for example, the project manager may keep a separate baseline for labor or procurements. The baseline may or may not include contingency funds or indirect costs. When earned value measurements are being used, the baseline may be called the performance measurement baseline.

The cost baseline can receive information from:

- Scope baseline
- Cost management plan
- Cost estimates
- Project schedule
- Resource management plan
- Risk register
- Agreements (contracts)

It provides information to:

- Project management plan
- Risk register

The cost baseline is an output from process 7.3 Determine Budget in the *PMBOK® Guide* – Sixth Edition. This is developed once and is not expected to change unless there is a significant change in scope.

Tailoring Tips

Consider the following tips to help tailor the cost baseline to meet your needs:

- The cost baseline generally includes contingency reserve to account for known risks. It does not generally include management reserve. Management reserve is held above the cost baseline, but is considered part of the project budget. If your company policies differ, the reserve in your baseline may be different.
- Your cost baseline may be displayed to show the funding constraints, funding requirements, or different sources of funding.
- Your organization may not require a graphic display of the cost baseline; they may require a spreadsheet or internal budgeting system displays instead.

Alignment

The cost baseline should be aligned and consistent with the following documents:

- Assumption log
- Project schedule
- Cost estimates
- Project team assignments
- Risk register

The cost baseline on the following page is displayed in a form called an S-curve.

COST BASELINE

Project Title: _____ **Date Prepared:** _____

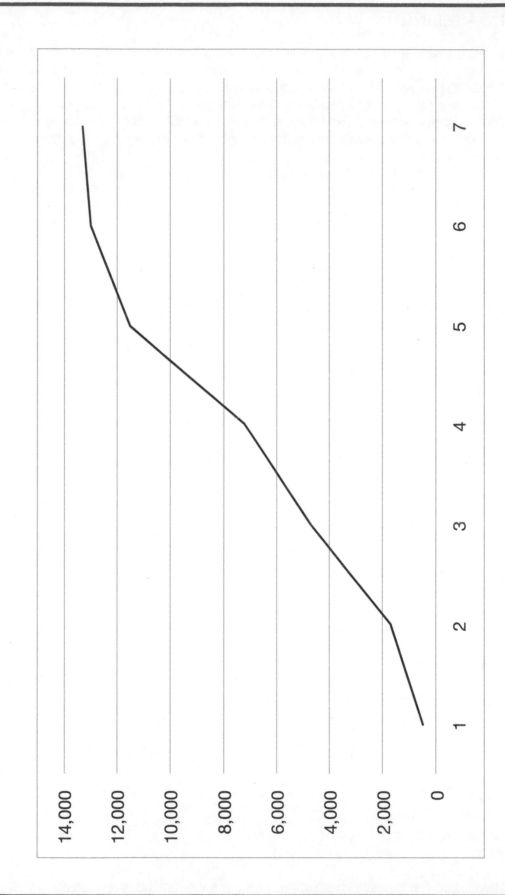

2.23 QUALITY MANAGEMENT PLAN

The quality management plan is a component of the project management plan. It describes how applicable policies, procedures, and guidelines will be implemented to achieve the quality objectives for the project. Information in the quality management plan can include:

- Quality standards that will be used on the project
- Quality objectives
- Quality roles and responsibilities
- Deliverables and processes subject to quality review
- Quality control and quality management activities for the project
- Quality procedures applicable for the project

The quality management plan can receive information from:

- Project charter
- Assumption log
- Stakeholder register
- Requirements management plan
- Risk management plan
- Stakeholder engagement plan
- Scope baseline
- Requirements documentation
- Requirements traceability matrix
- Risk register

It provides information to:

- Scope management plan
- Cost estimates
- Resource management plan
- Risk register
- Procurement documents (RFP, RFQ)

The quality management plan is an output from process 8.1 Plan Quality Management in the *PMBOK®️ Guide* – Sixth Edition. This is developed once and is not usually changed.

Tailoring Tips

Consider the following tips to help tailor the quality management plan to meet your needs:

- On smaller projects quality, requirements, and scope are often handled as a single aspect, whereas in larger projects they are separated out and may have distinct roles and responsibilities for each aspect.
- In many industries there are specific standards that must be adhered to. Your quality management plan may reference these by citing specific regulations, or they may be integrated into organizational policies and procedures.
- Quality management planning must be consistent with your organization's quality policies, processes, and procedures.

Alignment

The quality management plan should be aligned and consistent with the following documents:

- Project charter
- Scope management plan
- Requirements management plan
- Resource management plan
- Procurement documents (RFP, RFQ, etc.)

Description

You can use the element descriptions in Table 2.21 to assist you in developing a quality management plan.

TABLE 2.21   Elements of a Quality Management Plan

Document Element	Description
Quality standards	Quality standards are usually industry or product driven. They may be ISO standards, IEEE, or some other regulatory or industry body.
Quality objectives	Quality objectives are the measures that must be achieved by the project or product components to meet the stakeholder needs. Objectives are the target you want to achieve. You may have metrics or specifications that provide a quantifiable measurement of success.
Quality roles and responsibilities	Define the roles necessary to conduct quality activities on the project and the responsibilities associated with each.
Deliverables and processes subject to quality review	The key deliverables that have metrics or measures associated with quality objectives
	The processes used in the project that require verification or validation that they are being performed correctly, or in accordance with quality requirements or objectives
Quality management approach	The approach that will be used to manage the quality process. Includes the timing and content of project and product quality audits.
Quality control approach	The approach that will be used to measure the product and the project performance to ensure the product meets the quality objectives
Applicable quality procedures	Procedures that will be used for the project, such as • Nonconformance and rework • Corrective actions • Quality audits • Continuous improvement

QUALITY MANAGEMENT PLAN

Project Title: _____ **Date Prepared:** _____

Quality Standards

Quality Objectives

Metric or Specification	Measure
1.	1.
2.	2.
3.	3.
4.	4.

Quality Roles and Responsibilities

Roles	Responsibilities
1.	1.
2.	2.
3.	3.

QUALITY MANAGEMENT PLAN

Deliverables and Processes Subject to Quality Review

Deliverables	Processes

Quality Management Approach

Quality Control Approach

Applicable Quality Procedures

2.24 QUALITY METRICS

Quality metrics provide specific detailed measurements about a project or product attribute, and how it should be measured to verify compliance. Metrics are consulted in the manage quality process to ensure that the processes used will meet the metric. The deliverables or processes are measured in the control quality phase and compared to the metric to determine if the result is acceptable or if corrective action or rework is required.

Quality metrics can receive information from:

* Project management plan
* Requirements documentation
* Stakeholder register

Quality metrics are an output from process 8.1 Plan Quality Management in the *PMBOK® Guide – Sixth Edition*. They are generally determined as the requirements are developed. If requirements are stable, they will be developed once. If requirements are evolving or changing, they will evolve and change as well.

Tailoring Tips

Consider the following tips to help tailor the quality metrics to meet your needs:

* On smaller projects quality metrics, requirements, and specifications are considered the same thing. Different industries may use the term "specifications" rather than "metrics."
* In many industries there are specific standards that include metrics. These must be adhered to in your project. Your quality management plan may reference these by citing specific regulations, or they may be integrated into organizational policies and procedures.

Alignment

The quality metrics should be aligned and consistent with the following documents:

* Requirements documentation
* Quality management plan

Description

You can use the element descriptions in Table 2.22 to assist you in documenting quality metrics.

TABLE 2.22   Elements of Quality Metrics

Document Element	Description
ID	Unique identifier. This can be the WBS ID or activity ID number.
Item	Describe the attribute to be measured.
Metric	The specific, quantifiable measurement
Measurement method	The method of measuring, including any equipment or procedures

QUALITY METRICS

Project Title: _____ Date Prepared: _____

ID	Item	Metric	Measurement Method

2.25 RESPONSIBILITY ASSIGNMENT MATRIX

The responsibility assignment matrix (RAM) shows the intersection of work packages and resources. Generally, RAMs are used to show the different levels of participation on a work package by various team members rather than physical resources. RAMs can indicate different types of participation depending on the needs of the project. Some common types include:

- Accountable
- Responsible
- Consulted
- Resource
- Informed
- Sign-off

The RAM always should include a key that explains what each of the levels of participation entails. An example follows using a RACI chart, as demonstrated in the *PMBOK® Guide* – Sixth Edition. The needs of your project should determine the fields for the RAM you use.

The responsibility assignment matrix can receive information from:

- Scope baseline
- Requirements documentation
- Stakeholder register

It is a data representation tool that provides information to the resource management plan in process 9.1 Plan Resource Management in the *PMBOK® Guide* Sixth Edition. It is progressively elaborated as more information about the scope and the resource requirements is known.

Tailoring Tips

Consider the following tips to help tailor the RAM to meet your needs:

- Tailor the types of participation appropriate for your project. Some projects require "sign-off" of specific deliverables, whereas others use the term "approve."
- Determine the appropriate level to record information on the RAM. Large projects with multiple vendors and large deliverables often use the RAM as the intersection of the WBS and the OBS (organizational breakdown structure). Small projects may use it at the deliverable or activity level to help enter schedule information.

Alignment

The RAM should be aligned and consistent with the following documents:

- Work breakdown structure
- Requirements documentation
- Resource requirements
- Procurement documents (RFP, RFQ, etc.)

Description

You can use the element descriptions in Table 2.23 to assist you in developing a responsibility assignment matrix.

TABLE 2.23   Elements of a Responsibility Assignment Matrix

Document Element	Description
Work package	Name of the work package you are assigning resources to. The RAM can be used at the work package level, control account level, or activity level.
Resource	Identify the person, division, or organization that will be working on the project.

RESPONSIBILITY ASSIGNMENT MATRIX

Project Title: _____ Date Prepared: _____

	Person 1	Person 2	Person 3	Person 4	Etc.
Work package 1	R	C	A		
Work package 2		A		I	R
Work package 3		R	R	A	
Work package 4	A	R	I	C	
Work package 5	C	R	R		A
Work package 6	R		A	I	
Etc.	C	A		R	R

R = Responsible: The person performing the work.

C = Consult: The person who has information necessary to complete the work.

A = Accountable: The person who is answerable to the project manager that the work is done on time, meets requirements, and is acceptable.

I = Inform: This person should be notified when the work is complete.

2.26 RESOURCE MANAGEMENT PLAN

The resource management plan is part of the project management plan. It provides guidance on how team and physical resources should be allocated, managed, and released. Information in the resource management plan includes:

- Estimating methods used to identify the type, number, and skill level of team resources
- Information on how project team members will be acquired and released
- Roles and responsibilities associated with the project
- Project organizational chart
- Training requirements
- Rewards and recognition
- Team development
- Methods used to identify the type, amount, and grade of physical resources
- Information on how physical resources will be acquired
- Methods for managing physical resources, such as inventory, supply chain, and logistics

The resource management plan can receive information from:

- Project charter
- Quality management plan
- Scope baseline
- Project schedule
- Requirements documentation
- Risk register
- Stakeholder register

It provides information to:

- Project budget
- Resource requirements
- Resource breakdown structure
- Team performance assessments
- Communications management plan
- Risk register
- Procurement management plan

The resource management plan is an output from 9.1 Plan Resource Management in the *PMBOK®* *Guide* – Sixth Edition. It is generally developed once and does not change.

Tailoring Tips

Consider the following tips to help tailor the resource management plan to meet your needs:

- If you need to bring in outside contractors for the project you will need to include information on how to on-board them to the project. You will also need to consider how to ensure they have all the information they need, but no access to proprietary data. This may include a "non-disclosure agreement" or similar forms.
- For any team or physical resources that are acquired from outside the organization you will need to work with procurement policies for the organization and the project.

- Projects with large amounts of inventory, supplies, or material should either reference organizational policies regarding managing physical resources, or provide sufficient detail to ensure appropriate control.

Alignment

The resource management plan should be aligned and consistent with the following documents:

- Work breakdown structure
- Requirements documentation
- Quality management plan
- Procurement management plan

Description

You can use the element descriptions in Table 2.24 to assist you in developing a resource management plan.

TABLE 2.24 Elements of a Resource Management Plan

Document Element	Description
Team member identification	Methods used to identify the skill sets needed and the level of skill needed. This includes techniques to estimate the number of resources needed, such as information from past projects, parametric estimates, or industry standards.
Team member acquisition	Document how staff will be brought on to the project. Describe any differences between internal team members and contract team members with regard to on-boarding procedures.
Team member management	Document how team members will be managed and eventually released from the team. Management methods may vary depending on the relative authority of the project manager and whether team members are internal to the organization or contract staff. Team member release should include methods for knowledge transfer.
Project organizational chart	Create a hierarchy chart to show the project reporting and organizational structure.
Roles and responsibilities	Provide information on the following: **Role.** Identify the role or job title and a brief description of the role. **Authority.** Define the decision-making, approval, and influence levels for each role. Examples include alternative selection, conflict management, prioritizing, rewarding and penalizing, etc. **Responsibility.** Define the activities that each role carries out, such as job duties, processes involved, and the hand-offs to other roles. **Qualifications.** Describe any prerequisites, experience, licenses, seniority levels, or other qualifications necessary to fulfill the role. **Competencies.** Describe specific role or job skills and capacities required to complete the work. May include details on languages, technology, or other information necessary to complete the roles successfully.
Training requirements	Describe any required training on equipment, technology, or company processes. Include information on how and when training will be accomplished.
Rewards and recognition	Describe any reward and recognition processes and limitations.
Team development	Describe methods for developing individual team members and the team as a whole.
Physical resource identification	Methods used to identify the materials, equipment, and supplies needed to complete the work. This includes units of measure and techniques to estimate the amount of resources needed, such as information from past projects, parametric estimates, or industry standards.
Physical resource acquisition	Document how equipment, materials, and supplies will be acquired. This can include buy, lease, rent, or pull from inventory. In the event resources are acquired, ensure alignment with procurement management processes.
Physical resource management	Document how materials, equipment, and supplies will be managed to ensure they are available when needed. This can include appropriate inventory, supply chain, and logistics information.

RESOURCE MANAGEMENT PLAN

Project Title: _____ Date Prepared: _____

Team Member Identification and Estimates

Role	Number	Skill Level
1.	1.	1.
2.	2.	2.
3.	3.	3.
4.	4.	4.
5.	5.	5.
6.	6.	6.

Staff Acquisition   Staff Release:

Roles, Responsibilities, and Authority

Role	Responsibility	Authority
1.	1.	1.
2.	2.	2.
3.	3.	3.
4.	4.	4.
5.	5.	5.
6.	6.	6.

Project Organizational Structure

RESOURCE MANAGEMENT PLAN

Training Requirements

Rewards and Recognition

Team Development

Physical Resource Identification and Estimates

Resource	Amount	Grade
1.	1.	1.
2.	2.	2.
3.	3.	3.
4.	4.	4.
5.	5.	5.
6.	6.	6.

Resource Acquisition

Resource Management

2.27 TEAM CHARTER

The team charter is used to establish ground rules and guidelines for the team. It is particularly useful on virtual teams and teams that are comprised of members from different organizations. Using a team charter can help establish expectations and agreements on working effectively together. The contents of the team charter typically include:

- Team values and principles
- Meeting guidelines
- Communication guidelines
- Decision-making process
- Conflict resolution process
- Team agreements

The team charter is an output from 9.1 Plan Resource Management in the *PMBOK® Guide* – Sixth Edition. It is generally developed once and does not change; however, if there is substantial team member turnover, the team should periodically revisit the team charter and reaffirm or update it accordingly.

Tailoring Tips

Consider the following tips to help tailor the team charter to meet your needs:

- If you bring in contractors for key roles in the project you should include them in developing the team charter.
- If your organization has organizational values, make sure your team charter is aligned with the organizational values.
- International teams may need to spend more time developing this document as different cultures have different ways of making decisions and resolving conflicts.

Alignment

The team charter should be aligned and consistent with the following documents:

- Resource management plan

Description

You can use the element descriptions in Table 2.25 to assist you in developing a team charter.

TABLE 2.25   Elements of a Team Charter

Document Element	Description
Team values and principles	List values and principles that the team agrees to operate within. Examples include mutual respect, operating from fact not opinion, etc.
Meeting guidelines	Identify guidelines that will keep meetings productive. Examples include decision makers must be present, start on time, stick to the agenda, etc.
Communication guidelines	List guidelines used for effective communication. Examples include everyone voices their opinion, no dominating the conversation, no interrupting, not using inflammatory language, etc.
Decision-making process	Describe the process used to make decisions. Indicate the relative power of the project manager for decision making as well as any voting procedures. Also indicate the circumstances under which a decision can be revisited.
Conflict resolution process	Describe the process for managing conflict, when a conflict will be escalated, when it should be tabled for later discussion, etc.
Other agreements	List any other agreements or approaches to ensuring a collaborative and productive working relationship among team members.

TEAM CHARTER

Project Title: _____ Date Prepared: _____

Team Values and Principles:

1.
2.
3.
4.
5.

Meeting Guidelines:

1.
2.
3.
4.
5.

Communication Guidelines:

1.
2.
3.
4.
5.

Decision-Making Process:

TEAM CHARTER

Conflict Resolution Process:

Other Agreements:

Signature:

Date:

2.28 RESOURCE REQUIREMENTS

The resource requirements describe the type and quantity of resources needed to complete the project work. Resources include:

- People
- Equipment
- Material
- Supplies
- Locations (as needed)

Locations can include training rooms, testing sites, and so on.

The activity resource requirements can receive information from:

- Assumption log
- Resource management plan
- Scope baseline
- Activity list
- Activity attributes
- Cost estimates resource calendars
- Risk register

It provides information to:

- Duration estimating worksheet
- Project schedule
- Cost estimating worksheet
- Risk register
- Procurement management plan

The resource requirements form is an output from process 9.2 Estimate Activity Resources in the *PMBOK® Guide* – Sixth Edition. Resource requirements are based on the project scope. Therefore, if the scope is known and stable, the requirements should remain relatively stable. If the scope is evolving, the resource requirements will evolve as well. The resource requirements will become more detailed and more stable over time.

Tailoring Tips

Consider the following tips to help tailor the resource requirements to meet your needs:

- You may want to divide the form into two sections—one for team resources and one for physical resources. As an alternative you can have one section for internal resources and one section for contracted or purchased resources.
- Consider adding a column that includes the basis of estimates. This can include supporting detail such as:
  - Method used for estimating the quantities
  - Range of estimates
  - Confidence level of estimates
  - Constraints or risks associated with the resource
- For projects with large amounts of inventory, supplies, or material you may want to document support requirements, such as inventory, supply chain, and logistical requirements.

Alignment

The resource requirements should be aligned and consistent with the following documents:

- Project schedule
- Cost estimates
- Bid documents

Description

You can use the element descriptions in Table 2.26 to assist you in developing resource requirements.

TABLE 2.26   Elements of Resource Requirements

Document Element	Description
ID	Unique identifier
Type of resource	Indicate whether the resource is a team resource or physical resource. If physical, indicate if it is equipment, supplies, material, location, or some other form of resource.
Quantity	Document the number or quantity of the resource needed for the activity. Indicate the unit of measure used for estimating resources.
Assumptions	Enter assumptions associated with the resource, such as availability, certifications, etc.
Comments	Include information on basis of estimate, grade, competency, or other relevant information.

RESOURCE REQUIREMENTS

Project Title: _____   Date Prepared: _____

ID	Resource	Quantity	Assumptions	Comments

2.29 RESOURCE BREAKDOWN STRUCTURE

The resource breakdown structure is a hierarchical structure used to organize the resources by type and category. It can be shown as a hierarchical chart or as an outline.

The resource breakdown structure can receive information from:

- Assumption log
- Resource management plan
- Scope baseline
- Activity attributes

It provides information to:

- Duration estimates worksheet

The resource breakdown structure is an output from process 9.2 Estimate Activity Resources in the *PMBOK® Guide* – Sixth Edition. Resources are based on the project scope. Therefore, if the scope is known and stable, the requirements should remain relatively stable. If the scope is evolving, the resource requirements will evolve as well.

Tailoring Tips

Consider the following tips to help tailor the resource breakdown structure to meet your needs:

- For projects with many different types of team resources you may want to decompose the team branch further by including information on skill level, required certifications, location, or other information.
- For projects that have different locations you may want to organize the resource breakdown structure by geography.

Alignment

The resource breakdown structure should be aligned and consistent with the following documents:

- Resource management plan
- Resource requirements

RESOURCE BREAKDOWN STRUCTURE

Project Title: _____ **Date Prepared:** _____

1. Project

    1.1. People

        1.1.1. Quantity of Role 1

            1.1.1.1.   Quantity of Level 1

            1.1.1.2.   Quantity of Level 2

            1.1.1.3.   Quantity of Level 3

        1.1.2. Quantity of Role 2

    1.2. Equipment

        1.2.1. Quantity of Type 1

        1.2.2. Quantity of Type 2

    1.3. Materials

        1.3.1. Quantity of Material 1

            1.3.1.1.   Quantity of Grade 1

            1.3.1.2.   Quantity of Grade 2

    1.4. Supplies

        1.4.1. Quantity of Supply 1

        1.4.2. Quantity of Supply 2

    1.5. Locations

        1.5.1. Location 1

        1.5.2. Location 2

2.30 COMMUNICATIONS MANAGEMENT PLAN

The communications management plan is a component of the project management plan. It describes how project communications will be planned, structured, implemented, and monitored for effectiveness. Typical information includes:

- Stakeholder communication requirements
- Information
- Method or media
- Time frame and frequency
- Sender
- Communication assumptions and constraints
- Glossary of common terminology

In addition, the communications management plan can include resources, time, and budgets associated with communication activities, methods for addressing sensitive or proprietary information, and methods for updating the communications management plan.

The communications management plan can receive information from:

- Project charter
- Requirements documentation
- Resource management plan
- Stakeholder register
- Stakeholder engagement plan

It provides information to:

- Stakeholder register
- Stakeholder engagement plan

The communications management plan is an output from process 10.1 Plan Communications Management in the *PMBOK® Guide* – Sixth Edition. It is updated periodically throughout the project as stakeholders are added and leave the project and as communications needs emerge and shift.

Tailoring Tips

Consider the following tips to help tailor the communications management plan to meet your needs:

- When multiple organizations are working on a project there will be additional information needed, such as:
  - Person responsible for authorizing release of internal or confidential information
  - How different communication hardware, software, and technologies will be addressed to ensure information gets to everyone regardless of their communication infrastructure
- If you have a multinational team your plan will need to account for the business language, currency unit of measure, translation, and other factors required to ensure effective communication across multiple countries and cultures.
- On a project that has a significant communication component you will want to identify the resources allocated for communication activities, the time requirements, and the budget allocated.
- For projects with complex communication needs, include a flowchart of the sequence of communication events.

Alignment

The communications management plan should be aligned and consistent with the following documents:

- Project schedule
- Stakeholder register
- Stakeholder engagement plan

Description

You can use the element descriptions in Table 2.27 to assist you in developing the communications management plan.

TABLE 2.27   Elements of a Communications Management Plan

Document Element	Description
Stakeholder communication requirements	The people or the groups of people who need to receive project information and their specific requirements
Information	Describe the information to be communicated, including language, format, content, and level of detail.
Method or media	Describe how the information will be delivered; for example, email, meetings, web meetings, etc.
Time frame and frequency	List how often the information is to be provided and under what circumstances.
Sender	Insert the name of the person or the group that will provide the information.
Communication constraints or assumptions	List any assumptions or constraints. Constraints can include descriptions of proprietary, secure, or sensitive information and relevant restrictions for distribution.
Glossary of common terminology	List any terms or acronyms unique to the project or that are used in a unique way.

COMMUNICATIONS MANAGEMENT PLAN

Project Title: _____   **Date Prepared:** _____

Stakeholder	Information	Method	Timing or Frequency	Sender

Assumptions	Constraints

Glossary of Terms or Acronyms:

Attach relevant communication diagrams or flowcharts.

Page 1 of 1

2.31 RISK MANAGEMENT PLAN

The risk management plan is a component of the project management plan. It describes how risk management activities will be structured and performed for both threats and opportunities. Typical information includes:

- Risk strategy
- Methodology
- Roles and responsibilities for risk management
- Funding to identify, analyze, and respond to risk
- Frequency and timing for risk management activities
- Risk categories
- Stakeholder risk appetite
- Definitions of probability
- Definitions of impact by objective
- Probability and impact matrix template
- Methods to track and audit risk management activities
- Risk report formats

The risk management plan can receive information from:

- Project charter
- Project management plan (any component)
- Stakeholder register

It provides information to:

- Cost management plan
- Quality management plan
- Risk register
- Stakeholder engagement plan

The risk management plan is an input to all the other risk management processes. It describes the approach to all other risk management processes and provides key information needed to conduct those processes successfully.

The risk management plan is an output from process 11.1 Plan Risk Management in the *PMBOK® Guide* – Sixth Edition. It is developed once and does not usually change.

Tailoring Tips

Consider the following tips to help tailor the risk management plan to meet your needs:

- For a small, simple, or short-term project you can use a simplified risk register with a 3 × 3 probability and impact matrix. You would also include risk information in the project status report rather than a separate risk report.
- For larger, longer, and more complex projects you will want to develop a robust risk management process, including a more granular probability and impact matrix, quantitative assessments for the schedule and budget baselines, risk audits, and risk reports.
- Projects that are using an agile approach will address risk at the start of each iteration and during the retrospective.

Alignment

The risk management plan should be aligned and consistent with the following documents:

- Scope management plan
- Schedule management plan
- Cost management plan
- Quality management plan
- Resource management plan
- Procurement management plan
- Stakeholder engagement plan

Description

You can use the element descriptions in Table 2.28 to assist you in developing the risk management plan.

TABLE 2.28   Elements of a Risk Management Plan

Document Element	Description
Strategy	The general approach to managing risk on the project
Methodology	Describe the methodology or approach to the risk management. This includes any tools, approaches, or data sources that will be used.
Roles and responsibilities	Document the roles and responsibilities for various risk management activities.
Risk categories	Identify categorization groups used to sort and organize risks. These can be used to sort risks on the risk register or for a risk breakdown structure, if one is used.
Risk management funding	Document the funding needed to perform the various risk management activities, such as utilizing expert advice or transferring risks to a third party. Also establishes protocols for establishing, measuring, and allocating contingency and management reserves.
Frequency and timing	Determine the frequency of conducting formal risk management activities and the timing of any specific activities.
Stakeholder risk appetite	Identify the risk thresholds of the organization(s) and key stakeholders on the project with regard to each objective.
Risk tracking and audit	Document how risk activities will be recorded and how risk management processes will be audited.
Definitions of probability	Document how probability will be measured and defined. Include the scale used and the definition for each level in the probability scale. The probability definitions should reflect the stakeholder risk appetite. For example:
	Very high = there is an 80 percent probability or higher that the event will occur
	High = there is a 60–80 percent probability that the event will occur
	Medium = there is a 40–60 percent probability that the event will occur
	Low = there is a 20–40 percent probability that the event will occur
	Very low = there is a 1–20 percent probability that the event will occur
Definitions of impact by objective	Document how impact will be measured and defined for either the project as a whole or for each objective. The probability definitions should reflect the stakeholder risk appetite. Include the scale used and the definition for each level in the impact scale. For example:
	Cost Impacts:
	Very high = overrun of control account budget of >20 percent
	High = overrun of control account budget between 15–20 percent
	Medium = overrun of control account budget between 10–15 percent
	Low = overrun of control account budget between 5–10 percent
	Very low = overrun of control account budget of <5 percent
Probability and impact matrix	Describe the combinations of probability and impact that indicate a high risk, a medium risk, and a low risk and the scoring that will be used to prioritize risks. This can also include an assessment of urgency to indicate how soon the risk event is likely to occur.

RISK MANAGEMENT PLAN

Project Title: _____ Date Prepared: _____

Strategy

Methodology

Roles and Responsibilities

Role	Responsibility
1.	1.
2.	2.
3.	3.
4.	4.

Risk Categories

Risk Management Funding

Contingency Protocols

RISK MANAGEMENT PLAN

Project Title: _____ **Date Prepared:** _____

Frequency and Timing

Stakeholder Risk Tolerances

Risk Tracking and Audit

RISK MANAGEMENT PLAN

Definitions of Probability

Very High	
High	
Medium	
Low	
Very Low	

Definitions of Impact by Objective

	Scope	Quality	Time	Cost
Very High				
High				
Medium				
Low				
Very Low				

RISK MANAGEMENT PLAN

Probability and Impact Matrix

Very High					
High					
Medium					
Low					
Very Low					
	Very Low	Low	Medium	High	Very High

2.32 RISK REGISTER

The risk register captures the details of identified individual risks. It documents the results of risk analysis, risk response planning, response implementation, and current status. It is used to track information about identified risks over the course of the project. Typical information includes:

- Risk identifier
- Risk statement
- Risk owner
- Probability of occurring
- Impact on objectives if the risk occurs
- Risk score
- Response strategies
- Revised probability
- Revised impact
- Revised score
- Actions
- Status
- Comments

The risk register can receive information from anywhere in the project environment. Some documents that should be specifically reviewed for input include:

- Assumption log
- Issue log
- Lessons learned register
- Requirements management plan
- Requirements documentation
- Scope baseline
- Schedule management plan
- Duration estimates
- Schedule baseline
- Cost management plan
- Cost estimates
- Cost baseline
- Quality management plan
- Resource management plan
- Resource requirements
- Risk management plan
- Procurement documents
- Agreements
- Stakeholder register

The risk register provides information to:

- Scope statement
- Duration estimates
- Cost estimates
- Quality management plan
- Resource requirements

- Risk report
- Procurement management plan
- Stakeholder engagement plan
- Lessons learned register
- Project closeout

The risk register is an output from process 11.2 Identify Risks in the *PMBOK® Guide* – Sixth Edition. It is developed at the start of the project and is updated throughout the project.

Description

You can use the element descriptions in Table 2.29 to assist you in developing the risk register.

TABLE 2.29   Elements of a Risk Register

Document Element	Description
Risk ID	Enter a unique risk identifier.
Risk statement	Describe the risk event or condition. A risk statement is usually phrased as "EVENT may occur, causing IMPACT" or "If CONDITION exists, EVENT may occur, leading to EFFECT."
Risk owner	The person responsible for managing and tracking the risk
Probability	Determine the likelihood of the event or condition occurring.
Impact	Describe the impact on one or more of the project objectives.
Score	If you are using numeric scoring, multiply the probability times the impact to determine the risk score. If you are using relative scoring then combine the two scores (e.g., high-low or medium-high).
Response	Describe the planned response strategy to the risk or condition.
Revised probability	Determine the likelihood of the event or condition occurring after the response has been implemented.
Revised impact	Describe the impact once the response has been implemented.
Revised score	Enter the revised risk score once the response has been implemented.
Actions	Describe any actions that need to be taken to respond to the risk.
Status	Enter the status as open or closed.
Comments	Provide any comments or additional helpful information about the risk event or condition.

RISK REGISTER

Project Title: _____   **Date Prepared:** _____

ID	Risk Statement	Owner	Probability	Impact				Score	Response
				Scope	Quality	Schedule	Cost		

Revised Probability	Impact				Revised Score	Responsible Party	Actions	Status	Comments
	Scope	Quality	Schedule	Cost					

2.33 RISK REPORT

The risk report presents information on overall project risk and summarizes information on individual project risks. It provides information for each of the processes from identification of risks, through analysis, response planning and implementation, and monitoring risks. Typical information includes:

- Executive summary
- Description of overall project risk
- Description of individual project risks
- Quantitative analysis
- Reserve status
- Risk audit results (if applicable)

The risk report can receive information from anywhere in the project environment. Some documents that should be specifically reviewed for input include:

- Assumption log
- Issue log
- Lessons learned register
- Risk management plan
- Project performance reports
- Variance analysis
- Earned value status
- Risk audit
- Contractor status reports

The risk report provides information to:

- Lessons learned register
- Project closeout report

The risk report is an output from process 11.2 Identify Risks in the *PMBOK® Guide* – Sixth Edition. It is developed at the start of the project and is updated throughout the project.

Tailoring Tips

Consider the following tips to help tailor the risk report to meet your needs:

- For a small, simple, or short-term project you can summarize this information in the regular project status report rather than create a separate risk report.
- Many projects do not include a quantitative risk analysis; if yours does not, omit this information from the report.
- For larger, longer, and more complex projects you can tailor the quantitative risk analysis techniques used to those most appropriate to your project.
- For more robust risk reports include appendices that may include the full risk register and quantitative risk model input (probabilistic distributions, branch correlation groups).

Alignment

The risk report should be aligned and consistent with the following documents:

- Assumption log
- Issue register

- Project performance report
- Risk management plan
- Risk register

Description

You can use the element descriptions in Table 2.30 to assist you in developing the risk report.

TABLE 2.30   Elements of a Risk Report

Document Element	Description
Executive summary	A statement describing the overall project risk exposure and major individual risks affecting the project, along with the proposed responses for trends.
Overall project risk	Provide a description of the overall risk of the project, including: • High-level statement of trends • Significant drivers of overall risk • Recommended responses to overall risk

Individual project risks	Analyze and summarize information associated with individual project risks, including:
	
	• Number of risks in each box of the probability impact matrix
	
	• Key metrics
	
	  • Active risks
	
	  • Newly closed risks
	
	  • Risks distribution by category, objective, and score
	
	• Most-critical risks and changes since last report
	
	• Recommended responses to top risks

Quantitative analysis	Summarize the results of quantitative risk analysis, including: • Results from quantitative assessments (S-curve, tornado, etc.) • Probability of meeting key project objectives • Drivers of cost and schedule outcomes • Proposed responses
Reserve status	Describe the reserve status, such as reserve used, reserve remaining, and an assessment of the adequacy of the reserve.
Risk audit results (if applicable)	Summarize the results of a risk audit of the risk management processes.

RISK REPORT

Project Title: _____ Date: _____

Executive Summary

OVERALL PROJECT RISK

Overall Risk Status and Trends

Significant Drivers of Overall Risk	Recommended Responses

INDIVIDUAL PROJECT RISKS

Indicate the number of individual risks in each box below.

VH					
H					
M					
L					
VL					
	VL	L	M	H	VH

RISK REPORT

Metrics

Number of scope risks	
Number of schedule risks	
Number of cost risks	
Number of quality risks	
Number of very high probability risks	
Number of high probability risks	
Number of medium probability risks	
Number of active risks	
Newly closed risks	

Critical Risks

Top Risks	Responses
1.	1.
2.	2.
3.	3.
4.	4.
5.	5.

Changes to Critical Risks

RISK REPORT

Quantitative Analysis Summary

Probability of Meeting Objectives:

Scope	Schedule	Cost	Quality	Other

Range of Outcomes

Range of Schedule Outcomes	Range of Cost Outcomes

Key Drivers of Variances	Proposed Responses

Reserve Status

Total Cost Reserve	Used to Date	Used This Period	Remaining Reserve

Total Schedule Reserve	Used to Date	Used This Period	Remaining Reserve

RISK REPORT

Assessment of Reserve Adequacy

Risk Audit Summary

Summary of Risk Events	
Summary of Risk Management Processes	
Summary of Recommendations	

2.34 PROBABILITY AND IMPACT ASSESSMENT

Definitions for probability and impact are defined in the risk management plan. If your project does not have a risk management plan, you can develop a probability and impact assessment to record definitions for the likelihood of events occurring (probability), and the impact on the various project objectives if they do occur. It also has a key to assign an overall risk rating based on the probability and impact scores.

The probability and impact assessment can receive information from:

* Risk management plan
* Risk register

It provides information to the risk register.

The probability and impact assessment is a tool used in 11.3 Perform Qualitative Risk Analysis in the *PMBOK® Guide* – Sixth Edition. It is developed once and does not usually change.

Tailoring Tips

Consider the following tips to help tailor the probability and impact matrix to meet your needs:

* On smaller projects the impacts may be grouped together without distinguishing impact by objective.
* The matrix can be 3 × 3 for a small project, 5 × 5 for a medium project, and 10 × 10 for a complex or large project.
* To indicate the relative criticality of various objectives (usually scope, schedule, cost, and quality), you can include a tighter range of thresholds between levels. For example, if cost is a critical factor, consider a very low impact as a 2 percent variance, a low variance as a 4 percent impact, a medium variance as a 6 percent variance, a high variance as 8 percent, and a very high variance as a 10 percent variance. If the cost is more relaxed you might have a loose range such as: very low impact as a 5 percent variance, a low variance as a 10 percent impact, a medium variance as a 15 percent variance, a high variance as 20 percent, and a very high variance as a 25 percent variance.
* If there are other objectives that are important to the project, such as stakeholder satisfaction, you can incorporate them. Some organizations combine scope and quality into one objective.
* You can make the assessment more robust by including urgency information to indicate if a risk is imminent or in the distant future.

Alignment

The probability and impact assessment should be aligned and consistent with the following documents:

* Risk management plan
* Risk register
* Probability and impact matrix
* Stakeholder register

Description

You can use the assessment descriptions in Table 2.31 to assist you in developing a probability and impact assessment.

TABLE 2.31   Elements of a Probability Impact Assessment

Document Element	Description	
Scope impact	Very High	The product does not meet the objectives and is effectively useless
	High	The product is deficient in multiple essential requirements
	Medium	The product is deficient in one major requirement or multiple minor requirements
	Low	The product is deficient in a few minor requirements
	Very Low	Minimal deviation from requirements
Quality impact	Very High	Performance is significantly below objectives and is effectively useless
	High	Major aspects of performance do not meet requirements
	Medium	At least one performance requirement is significantly deficient
	Low	There is minor deviation in performance
	Very Low	There is minimal deviation in performance
Schedule impact	Very High	Greater than 20 percent overall schedule increase
	High	Between 10 percent and 20 percent overall schedule increase
	Medium	Between 5 percent and 10 percent overall schedule increase
	Low	Noncritical paths have used all their float, or overall schedule increase of 1 to 5 percent
	Very Low	Slippage on noncritical paths but float remains
Cost impact	Very High	Cost increase of greater than 20 percent
	High	Cost increase of 10 to 20 percent
	Medium	Cost increase of 5 to 10 percent
	Low	Cost increase that requires use of all contingency funds
	Very Low	Cost increase that requires use of some contingency but some contingency funds remain
Probability	Very High	The event will most likely occur: 80 percent or greater probability
	High	The event will probably occur: 61 to 80 percent probability
	Medium	The event is likely to occur: 41 to 60 percent probability
	Low	The event may occur: 21 to 40 percent probability
	Very Low	The event is unlikely to occur: 1 to 20 percent probability
Risk rating	High	Any event with a probability of medium or above and a very high impact on any objective Any event with a probability of high or above and a high impact on any objective Any event with a probability of very high and a medium impact on any objective Any event that scores a medium on more than two objectives

TABLE 2.31   Elements of a Probability Impact Assessment (*continued*)

Document Element	Description	
	Medium	Any event with a probability of very low and a high or above impact on any objective
		Any event with a probability of low and a medium or above impact on any objective
		Any event with a probability of medium and a low to high impact on any objective
		Any event with a probability of high and a very low to medium impact on any objective
		Any event with a probability of very high and a low or very low impact on any objective
		Any event with a probability of very low and a medium impact on more than two objectives
	Low	Any event with a probability of medium and a very low impact on any objective
		Any event with a probability of low and a low or very low impact on any objective
		Any event with a probability of very low and a medium or less impact on any objective

PROBABILITY AND IMPACT RISK RATING

Project Title: _____ **Date Prepared:** _____

Scope Impact

Very High	
High	
Medium	
Low	
Very Low	

Quality Impact

Very High	
High	
Medium	
Low	
Very Low	

Schedule Impact

Very High	
High	
Medium	
Low	
Very Low	

PROBABILITY AND IMPACT RISK RATING

Cost Impact

Very High	
High	
Medium	
Low	
Very Low	

Probability

Very High	
High	
Medium	
Low	
Very Low	

Risk Rating

High	
Medium	
Low	

2.35 PROBABILITY AND IMPACT MATRIX

The probability and impact matrix is a table that is used to plot each risk after performing a probability and impact assessment. The probability and impact assessment determines the probability and impact of the risk. This matrix provides a helpful way to view the various risks on the project and prioritize them for responses. It may be constructed for threats and opportunities. Information from this matrix will be transferred to the risk register.

This matrix also provides an overview of the amount of risk on the project. The project team can get an idea of the overall project risk by seeing the number of risks in each square of the matrix. A project with many risks in the red zone will need more contingency to absorb the risk and likely more time and budget to develop and implement risk responses.

The probability and impact matrix can receive information from:

- Risk management plan
- Probability and impact risk assessment
- Risk register

It provides information to the risk register.

The probability and impact matrix is a tool used in 11.3 Perform Qualitative Risk Analysis in the *PMBOK® Guide* – Sixth Edition. It is updated throughout the project.

Tailoring Tips

Consider the following tips to help tailor the probability and impact matrix to meet your needs:

- The matrix can be a 3 × 3 for a small project, 5 × 5 for a medium project, and 10 × 10 for a complex or large project.
- The numbering structure of the probability and impact matrix can be tailored to emphasize the high risks by creating a nonlinear numbering structure. For example, impact scores can be set up to double every increment. For example: Very Low = .5, Low = 1, Medium = 2, High = 4, and Very High = 8.
- The relative importance of the objectives can be weighted. If schedule is most important you may weight that as 40 percent of the score, whereas scope, quality, and cost all have a weight of 20 percent (make sure the total is 100 percent).
- The combination of probability and impact that indicates a risk is high, medium, or low can be tailored to reflect the organization's risk appetite. An organization with a low risk appetite may rank events that fall in the medium or high range for both impact and probability as high risk. An organization with a higher risk threshold may only rank risk with a very high probability and impact as high risk.

Alignment

The probability and impact assessment and matrix should be aligned and consistent with the following documents:

- Risk management plan
- Risk register
- Probability and impact assessment
- Stakeholder register

PROBABILITY AND IMPACT MATRIX

Project Title: _____   Date Prepared: _____

	Very Low	Low	Medium	High	Very High
Very High					
High					
Medium					
Low					
Very Low					

2.36 RISK DATA SHEET

A risk data sheet contains information about a specific identified risk. The information is filled in from the risk register and updated with more detailed information. Typical information includes:

- Risk identifier
- Risk description
- Status
- Risk cause
- Probability
- Impact on each objective
- Risk score
- Response strategies
- Revised probability
- Revised impact
- Revised score
- Responsible party
- Actions
- Secondary risks
- Residual risks
- Contingency plans
- Schedule or cost contingency
- Fallback plans
- Comments

The risk data sheet can receive information from:

- Risk register

It can be started in process 11.2 Identify Risks from the *PMBOK® Guide* – Sixth Edition. It is continuously updated and elaborated throughout the project.

Tailoring Tips

Consider the following tips to help tailor the risk data sheet to meet your needs:

- Not all projects require risk data sheets. Where they are used, they are considered an extension of the risk register.
- You can add or delete any fields you feel necessary.

Alignment

The risk data sheet should be aligned and consistent with the following documents:

- Risk register
- Probability and impact matrix
- Risk report

Description

You can use the element descriptions in Table 2.32 to assist you in developing the risk data sheet.

TABLE 2.32   Elements of a Risk Data Sheet

Document Element	Description
Risk ID	Enter a unique risk identifier.
Risk description	Provide a detailed description of the risk.
Status	Enter the status as open or closed.
Risk cause	Describe the circumstances or drivers that are the source of the risk.
Probability	Determine the likelihood of the event or condition occurring.
Impact	Describe the impact on one or more of the project objectives.
Score	If you are using numeric scoring, multiply the probability times the impact to determine the risk score. If you are using relative scoring then combine the two scores (e.g., high-low or medium-high).
Reponses	Describe the planned response strategy to the risk or condition.
Revised probability	Determine the likelihood of the event or condition occurring after the response has been implemented.
Revised impact	Describe the impact once the response has been implemented.
Revised score	Enter the revised risk score once the response has been implemented.
Responsible party	Identify the person responsible for managing the risk.
Actions	Describe any actions that need to be taken to respond to the risk.
Secondary risks	Describe new risks that arise out of the response strategies taken to address the risk.
Residual risk	Describe the remaining risk after response strategies.
Contingency plan	Develop a plan that will be initiated if specific events occur, such as missing an intermediate milestone. Contingency plans are used when the risk or residual risk is accepted.
Contingency funds	Determine the funds needed to protect the budget from overrun.
Contingency time	Determine the time needed to protect the schedule from overrun.
Fallback plans	Devise a plan to use if other response strategies fail.
Comments	Provide any comments or additional helpful information about the risk event or condition.

RISK DATA SHEET

Project Title: _____ Date Prepared: _____

Risk ID:

Risk Description:

Status:

Risk Cause:

Probability	Impact				Responses	Score
	Scope	Quality	Schedule	Cost		

Revised Probability	Revised Impact				Responsible Party	Revised Score
	Scope	Quality	Schedule	Cost		

Actions

Secondary Risks:

Residual Risk:

Contingency Plan:

Contingency Funds:

Contingency Time:

Fallback Plans:

Comments:

2.37 PROCUREMENT MANAGEMENT PLAN

The procurement management plan is a component of the project management plan that describes the activities undertaken during the procurement process. It describes how all aspects of a procurement will be managed. Typical information includes:

- How procured work will be coordinated and integrated with other project work. Of specific interest is:
  - Scope
  - Schedule
  - Documentation
  - Risk
- Timing of procurement activities
- Contract performance metrics
- Procurement roles, responsibility, and authority
- Procurement constraints and assumptions
- Legal jurisdiction and currency
- Information on use of independent estimates
- Risk management concerns, such as need for bonds or insurance
- Prequalified sellers lists

The procurement management plan can receive information from:

- Project charter
- Stakeholder register
- Scope management plan
- Requirements documentation
- Requirements traceability matrix
- Scope baseline
- Milestone list
- Project schedule
- Quality management plan
- Resource management plan
- Resource requirements
- Risk register
- Project team assignments

It provides information to:

- Risk register
- Stakeholder register

The procurement management plan is an output from process 12.1 Plan Procurement Management in the *PMBOK® Guide* – Sixth Edition. It is developed once and does not usually change.

Tailoring Tips

Consider the following tips to help tailor the procurement management plan to meet your needs:

- For a project that will be done using internal resources only, you do not need a procurement management plan.

- For a project where materials will be purchased and there is a standing purchase order with a vendor, you will not need a procurement management plan.
- For projects with a few procurements consider combining this form with the procurement strategy.
- You may wish to combine the assumptions and constraints for procurements with the assumption log.
- Work with the contracting or legal department to ensure compliance with organizational purchasing policies.

Alignment

The procurement management plan should be aligned and consistent with the following documents:

- Scope management plan
- Requirements management plan
- Scope baseline
- Schedule management plan
- Project schedule
- Cost management plan
- Cost estimates
- Project budget
- Procurement strategy

Description

You can use the element descriptions in Table 2.33 to assist you in developing the procurement management plan.

TABLE 2.33 Elements of a Procurement Management Plan

Document Element	Description	
Procurement integration	Scope	Define how the contractor's WBS will integrate with the project WBS.
	Schedule	Define how the contractor's schedule will integrate with the project schedule, including milestones and long lead items.
	Documentation	Describe how contractor documentation will integrate with project documentation.
	Risk	Describe how risk identification, analysis, and response will integrate with risk management for the overall project.
	Reporting	Define how the contractor's status reports will integrate with the project status report.
Timing	Identify the timetable of key procurement activities. Examples include when the statement of work (SOW) will be complete, when procurement documents will be released, the date proposals are due, and so forth.	
Performance metrics	Document the metrics that will be used to evaluate the seller's performance.	
Roles, responsibilities, and authority	Define the roles, responsibilities, and authority level of the project manager, contractor, and procurement department, as well as any other significant stakeholders for the contract.	
Assumptions and constraints	Record assumptions and constraints related to the procurement activities.	
Legal jurisdiction and currency	Identify the location that has legal jurisdiction. Identify the currency that will be used for pricing and payment.	
Independent estimates	Document whether independent cost estimates will be used and if they will be needed for source selection.	
Risk management	Document requirements for performance bonds or insurance contracts to reduce risk.	
Prequalified sellers	List any prequalified sellers that will be used.	

PROCUREMENT MANAGEMENT PLAN

Project Title: _____ Date: _____

Procurement Integration

Area	Integration Approach
Scope	
Schedule	
Documentation	
Risk	
Reporting	

Timing of Key Procurement Activities

Date	Activity

Performance Metrics

Item	Metric	Measurement Method

PROCUREMENT MANAGEMENT PLAN

Roles, Responsibility, and Authority

Role	Responsibility	Authority

Assumptions and Constraints

Category	Assumption/Constraint

Legal Jurisdiction and Currency

Independent Estimates

Risk Management

Prequalified Sellers

1.
2.
3.
4.

2.38 PROCUREMENT STRATEGY

The procurement strategy is a project document that describes information about specific procurements. Typical information includes:

- Delivery methods
- Contract types
- Procurement phases

The procurement strategy can receive information from:

- Project charter
- Stakeholder register
- Project roadmap
- Requirements documentation
- Requirements traceability matrix
- Scope baseline
- Milestone list
- Project schedule
- Resource management plan
- Resource requirements

It provides information to:

- Project schedule
- Project budget
- Quality management plan
- Risk register

The procurement strategy is an output from process 12.1 Plan Procurement Management in the *PMBOK® Guide* – Sixth Edition. It is developed once for each procurement when needed.

Tailoring Tips

Consider the following tips to help tailor the procurement strategy to meet your needs:

- For a project that will be done using internal resources only, you do not need a procurement strategy.
- For projects with few procurements consider combining this form with the procurement management plan.
- For simple purchases, or for purchases where you have worked with a vendor successfully for a length of time, you may not need a formal procurement strategy; rather you would record the information in a statement of work (SOW) that would be part of the contract.
- Work with the contracting or legal department to ensure compliance with organizational purchasing policies.

Alignment

The procurement strategy should be aligned and consistent with the following documents:

- Project charter
- Project roadmap

- Requirements documentation
- Requirements traceability matrix
- Schedule management plan
- Cost management plan
- Resource management plan
- Procurement management plan

Description

You can use the element descriptions in Table 2.34 to assist you in developing the procurement strategy.

TABLE 2.34 Elements of a Procurement Strategy

Document Element	Description	
Delivery methods	Professional services	Describe how the contractor will work with the buyer; for example, in a joint venture, as a representative, with or without subcontracting allowed.
	Construction services	Describe the limitations of delivery, such as design build, design bid build, etc.
Contract types	Describe the contract type, fixed, incentive, or award fees. Include the criteria associated with the fees. Common contract types Include:	
	Fixed Price:	
	FFP – Firm Fixed Price FPIF – Fixed Price with Incentive Fee FP-EPA – Fixed Price with Economic Price Adjustment	
	Cost Reimbursable:	
	CPFF – Cost Plus Fixed Fee CPIF – Cost Plus Incentive Fee CPAF – Cost Plus Award Fee	
	Time and Materials (T&M)	
Procurement phases	List the procurement phases, milestones, criteria to advance to the next phase, and tests or evaluations for each phase. Include any knowledge transfer requirements.	

PROCUREMENT STRATEGY

Project Title: _____ **Date:** _____

Delivery Method

Contract Type

☐ FFP ☐ FPIF ☐ FP-EPA ☐ CPFF ☐ CPIF ☐ CPAF ☐ T&M ☐ Other

Incentive or Award Fee	Criteria		

Procurement Life Cycle

Phase	Entry Criteria	Key Deliverables or Milestones	Exit Criteria	Knowledge Transfer

2.39 SOURCE SELECTION CRITERIA

Source selection criteria is a set of attributes desired by the buyer that a seller must meet or exceed to be selected for a contract. The source selection criteria form is an aid in determining and rating the criteria that will be used to evaluate bid proposals. This is a multistep process.

1. The criteria to evaluate bid responses are determined.
2. A weight is assigned to each criterion. The sum of all the criteria must equal 100 percent.
3. The range of ratings for each criterion is determined, such as 1–5 or 1–10.
4. The performance necessary for each rating is defined.
5. Each proposal is evaluated against the criteria and is rated accordingly.
6. The weight is multiplied by the rate and a score for each criterion is derived.
7. The scores are totaled and the highest total score is the winner of the bid.

Evaluation criteria commonly include such items as:

- Capability and capacity
- Product cost and life cycle cost
- Delivery dates
- Technical expertise
- Prior experience
- Proposed approach and work plan
- Key staff qualifications, availability, and competence
- Financial stability
- Management experience
- Training and knowledge transfer

The source selection criteria is an output from process 12.1 Plan Procurement Management in the *PMBOK® Guide* – Sixth Edition. Source selection criteria is established for each major procurement.

Tailoring Tips

Consider the following tips to help tailor the source selection criteria to meet your needs:

- For a small procurement that is not complex, you likely won't need to develop a weighted source selection criteria form.
- For international procurements you may also want to include familiarity with local laws and regulations, as well as experience and relationships in the locations involved.
- For construction projects, or projects with many logistical concerns, you could include logistics handling as a selection criteria.

Alignment

The source selection criteria should be aligned and consistent with the following documents:

- Requirements documentation
- Scope baseline
- Project schedule
- Resource management plan

- Resource requirements
- Procurement management plan
- Procurement strategy

Description

You can use the element descriptions in Table 2.35 to assist you in developing the source selection criteria.

TABLE 2.35   Elements of Source Selection Criteria

Document Element	Description	
Criteria	1	Describe what a 1 means for the criteria. For example, for experience, it may mean that the bidder has no prior experience.
	2	Describe what a 2 means for the criteria. For example, for experience, it may mean that the bidder has done 1 similar job.
	3	Describe what a 3 means for the criteria. For example, for experience, it may mean that the bidder has done 3 to 5 similar jobs.
	4	Describe what a 4 means for the criteria. For example, for experience, it may mean that the bidder has done 5 to 10 similar jobs.
	5	Describe what a 5 means for the criteria. For example, for experience, it may mean that the job is the bidder's core competency.
Weight	Enter the weight for each criterion. Total weight for all criteria must equal 100 percent.	
Candidate rating	Enter the rating per the criteria above.	
Candidate score	Multiply the weight times the rating.	
Total	Sum the scores for each candidate.	

SOURCE SELECTION CRITERIA

Project Title: _____   Date Prepared: _____

	1	2	3	4	5
Criterion 1					
Criterion 2					
Criterion 3					
Criterion 4					
Criterion 5					

	Weight	Candidate 1 Rating	Candidate 1 Score	Candidate 2 Rating	Candidate 2 Score	Candidate 3 Rating	Candidate 3 Score
Criterion 1							
Criterion 2							
Criterion 3							
Criterion 4							
Criterion 5							
Totals							

2.40 STAKEHOLDER ENGAGEMENT PLAN

The stakeholder engagement plan is a component of the project management plan. It describes the strategies and actions that will be used to promote productive involvement of stakeholders in decision making and project execution. Typical information includes:

- Desired and current engagement level of key stakeholders
- Scope and impact of change to stakeholders
- Interrelationships and potential overlap between stakeholders
- Engagement approach for each stakeholder or group of stakeholders

In addition, the stakeholder engagement plan can include methods for updating and refining the plan throughout the project.

The stakeholder engagement plan can receive information from:

- Project charter
- Assumption log
- Change log
- Issue log
- Stakeholder register
- Resource management plan
- Project schedule
- Communications management plan
- Risk management plan
- Risk register

It provides information to:

- Requirements documentation
- Quality management plan
- Communications management plan
- Stakeholder register

The stakeholder engagement plan is an output from process 13.2 Plan Stakeholder Engagement in the *PMBOK® Guide* – Sixth Edition. It is updated periodically throughout the project as needed.

Tailoring Tips

Consider the following tips to help tailor the stakeholder engagement plan to meet your needs:

- For small projects you may not need a stakeholder engagement plan. You can combine the information with the stakeholder register as appropriate.
- Projects with multiple stakeholders with overlapping and intersecting relationships can benefit from having a stakeholder relationship map that shows the connections.
- For many high-risk projects, stakeholder engagement is critical to success. For projects with many stakeholders, complex interactions, and high-risk stakeholders you will need to have a robust stakeholder engagement plan.

Alignment

The stakeholder engagement plan should be aligned and consistent with the following documents:

- Stakeholder register
- Communications management plan
- Project schedule

Description

You can use the element descriptions in Table 2.36 to assist you in developing the stakeholder engagement plan.

TABLE 2.36   Elements of a Stakeholder Engagement Plan

Document Element	Description
Stakeholder engagement assessment matrix	Use information from the stakeholder register to document stakeholders. Document "current" stakeholder engagement level with a "C" and "desired" stakeholder engagement with a "D." A common format includes the following stakeholder participation descriptions: **Unaware.** Unaware of project and its potential impacts **Resistant.** Aware of project and potential impacts and resistant to the change **Neutral.** Aware of project yet neither supportive nor resistant **Supportive.** Aware of project and potential impacts and supportive of change **Leading.** Aware of project and potential impacts and actively engaged in ensuring project success
Stakeholder changes	Describe any pending additions, deletions, or changes to stakeholders and the potential impact to the project.
Interrelationships	List any relationships between and among stakeholder groups.
Stakeholder engagement approach	Describe the approach you will use with each stakeholder to move them to the preferred level of engagement.

STAKEHOLDER ENGAGEMENT PLAN

Project Title: _____   **Date Prepared:** _____

Stakeholder	Unaware	Resistant	Neutral	Supportive	Leading

C = Current level of engagement   D = Desired level of engagement

Pending Stakeholder Changes

Stakeholder Relationships

STAKEHOLDER ENGAGEMENT PLAN

Stakeholder Engagement Approach

Stakeholder	Approach							

Executing Forms

3

3.0 EXECUTING PROCESS GROUP

The purpose of the Executing Process Group is to carry out the work necessary to meet the project objectives. There are ten processes in the Executing Process Group.

- Direct and Manage Project Work
- Manage Project Knowledge
- Manage Quality
- Acquire Resources
- Develop Project Team
- Manage Project Team
- Manage Communications
- Implement Risk Responses
- Conduct Procurements
- Manage Stakeholder Engagement

The intent of the Executing Process Group is to at least:

- Create the deliverables
- Manage the project knowledge including lessons learned
- Manage project quality
- Acquire team and physical resources
- Manage the project team
- Carry out project communications
- Implement risk responses
- Report project progress
- Bid and award contracts
- Manage stakeholder engagement with the project

In these processes, the main work of the project is carried out and the majority of the funds are expended. To be effective, the project manager must make decisions, identify issues, coordinate project resources, report progress, and engage with stakeholders, while completing the project deliverables.

The forms used to document project execution include:

- Issue log
- Decision log
- Change request
- Change log
- Lessons learned register
- Quality audit*
- Form test and evaluation documents
- Team performance assessment

---

*Test and evaluation documents, metrics, and measurements are product specific. There is no template that can provide an accurate depiction. They can include customized checklists, requirements traceability matrixes, performance requirements, and other product-specific information.

3.1 ISSUE LOG

The issue log is a project document used to record and monitor issues. An issue is defined as a current condition or situation that could have an impact on the project objectives. Examples of issues are points or matters in question that are in dispute or under discussion, or over which there are opposing views or disagreements. Issues can also arise from a risk event that has occurred and must now be dealt with. An issue log includes:

- Identifier
- Type
- Issue description
- Priority
- Impact on objectives
- Responsible party
- Status
- Resolution date
- Final resolution
- Comments

The issue log is an output from process 4.3 Direct and Manage Project Work in the *PMBOK® Guide –* Sixth Edition. It is a dynamic document that is created at the start of the project and is maintained throughout the project.

Tailoring Tips

Consider the following tips to help tailor the issue log to meet your needs:

- You may want to add information on the source of the issue.
- You could add a field that documents which stakeholders are impacted by the issue or should be involved with resolving the issue.

Alignment

The issue log should be aligned and consistent with the following documents:

- Risk register
- Decision log
- Lessons learned register

Description

You can use the element descriptions in Table 3.1 to assist you in developing the issue log.

TABLE 3.1   Elements of an Issue Log

Document Element	Description
ID	Enter a unique issue identifier.
Type	Document the type or category of the issue, such as stakeholder issue, technical issue, conflict, etc.
Issue description	Provide a detailed description of the issue.
Priority	Define the priority, such as urgent, high, medium, or low.
Impact on objectives	Identify the project objectives that the issue impacts and the degree of impact.
Responsible party	Identify the person who is assigned to resolve the issue.
Status	Denote the status of the issue as open or closed.
Resolution date	Document the date by which the issue needs to be resolved.
Final resolution	Describe how the issue was resolved.
Comments	Document any clarifying comments about the issue, resolution, or other fields on the form.

ISSUE LOG

Project Title: _____ Date Prepared: _____

Issue ID	Type	Issue Description	Priority	Impact on Objectives

Responsible Party	Status	Res. Date	Final Resolution	Comments

3.2 DECISION LOG

Frequently there are alternatives in developing a product or managing a project. Using a decision log can help keep track of the decisions that were made, who made them, and when they were made. A decision log can include:

- Identifier
- Category
- Decision
- Responsible party
- Date
- Comments

The decision log is not listed explicitly in the *PMBOK® Guide* – Sixth Edition; however, it can be very helpful in managing the day-to-day activities of the project. It is a dynamic document that is created at the start of the project and is maintained throughout the project.

Tailoring Tips

Consider the following tips to help tailor the decision log to meet your needs:

- For projects that are large, complicated, or complex you can add fields to identify the impacts of the decision on deliverables or project objectives.
- You could add a field that documents which stakeholders are impacted by the decision or should be involved with making the decision or should be informed of the decision.

Alignment

The decision log should be aligned and consistent with the following documents:

- Project scope statement
- Responsibility assignment matrix
- Communications management plan
- Issue register

Description

You can use the element descriptions in Table 3.2 to assist you in developing the decision log.

TABLE 3.2   Elements of a Decision Log

Document Element	Description
ID	Enter a unique decision identifier.
Category	Document the type of decision, such as technical, project, process, etc.
Decision	Provide a detailed description of the decision.
Responsible party	Identify the person authorized to make the decision.
Priority	Enter the date the decision was made and authorized.
Comments	Enter any further information to clarify the decision, alternatives considered, the reason the decision was made, and the impact of the decision.

DECISION LOG

Project Title: _____ Date Prepared: _____

ID	Category	Decision	Responsible Party	Date	Comments

3.3 CHANGE REQUEST

A change request is used to change any aspect of the project. It can pertain to product, documents, cost, schedule, or any other aspect of the project. Typical information includes:

- Requestor
- Category
- Description of the proposed change
- Justification
- Impacts of the proposed change
  - Scope
  - Quality
  - Requirements
  - Cost
  - Schedule
  - Project documents
- Comments

A change request can come from almost any process involved in executing, monitoring, and controlling the project. It is described in 4.3 Direct and Manage Project Work in the *PMBOK® Guide* – Sixth Edition. Upon completion, it is submitted to the change control board for review.

Tailoring Tips

Consider the following tips to help tailor the change request to meet your needs:

- For smaller projects you can simplify the form by having a summary description of the impacts without including impacts for each subcategory (scope, quality, requirements, etc.).
- You can add a check box that indicates whether the change is mandatory (such as a legal requirement) or discretionary.
- A field can be added that describes the implications of not making the change.

Alignment

The change request should be aligned and consistent with the following documents:

- Change management plan
- Change log

Description

You can use the element descriptions in Table 3.3 to assist you in developing the change request.

TABLE 3.3   Elements of a Change Request

Document Element	Description		
Requestor	The name, and if appropriate, the position of the person requesting the change		
Category	Check a box to indicate the category of change.		
Description of change	Describe the proposed change in enough detail to clearly communicate all aspects of the change.		
Justification for proposed change	Indicate the reason for the change.		
Impacts of change	Scope	Describe the impact of the proposed change on the project and product scope.	
	Quality	Describe the impact of the proposed change on the project or product quality.	
	Requirements	Describe the impact of the proposed change on the project or product requirements.	
	Cost	Describe the impact of the proposed change on the project budget, cost estimates, or funding requirements.	
	Schedule	Describe the impact of the proposed change on the schedule and whether it will change the critical path.	
	Project documents	Describe the impact of the proposed change on each project document.	
Comments	Provide any comments that will clarify information about the requested change.		

CHANGE REQUEST

Project Title: _____ Date Prepared: _____

Requestor:

Category:

☐ Scope   ☐ Quality   ☐ Requirements

☐ Cost   ☐ Schedule   ☐ Documents

Detailed Description of Proposed Change

Justification for Proposed Change

Impacts of Change

Scope	☐ Increase	☐ Decrease	☐ Modify
Description:			
Quality	☐ Increase	☐ Decrease	☐ Modify
Description:			

CHANGE REQUEST

Requirements	☐ Increase	☐ Decrease	☐ Modify

Description:

Cost	☐ Increase	☐ Decrease	☐ Modify

Description:

Schedule	☐ Increase	☐ Decrease	☐ Modify

Description:

Stakeholder Impact	☐ High risk	☐ Low risk	☐ Medium risk

Description:

Project Documents

Comments

CHANGE REQUEST

Disposition: ☐ Approve ☐ Defer ☐ Reject

Justification

3.4 CHANGE LOG

The change log is used to track changes from the change request through the final decision. Typical information includes:

- Identifier
- Category
- Description
- Requestor
- Submission date
- Status
- Disposition

The change log is related to the:

- Change request
- Change management plan

The change log is used in 4.6 Perform Integrated Change Control in the *PMBOK® Guide* – Sixth Edition. It is a dynamic document that is updated throughout the project.

Tailoring Tips

Consider the following tips to help tailor the change log to meet your needs:

- You can include additional summary information from the change request in the log, such as cost or schedule impact.
- You can add a check box that indicates whether the change is mandatory (such as a legal requirement) or discretionary.
- The change log can also record information to track configuration management, such as which configurable items are impacted.
- Some IT projects include a field that indicates if a change is a bug fix.

Alignment

The change log should be aligned and consistent with the following documents:

- Change management plan
- Change request

Description

You can use the element descriptions in Table 3.4 to assist you in developing the change log.

TABLE 3.4   Elements of a Change Log

Document Element	Description
Identifier	Enter a unique change identifier.
Category	Enter the category from the change request form.
Description	Describe the proposed change.
Requestor	Enter the name of the person requesting the change.
Submission date	Enter the date the change was submitted.
Status	Enter the status as open, pending, closed.
Disposition	Enter the outcome of the change request as approved, deferred, or rejected.

CHANGE LOG

Project Title: _____ Date Prepared: _____

ID	Category	Description of Change	Requestor	Submission Date	Status	Disposition

Page 1 of 1

3.5 LESSONS LEARNED REGISTER

The lessons learned register is used to record challenges, problems, good practices, and other information that can be passed along to the organization and to other projects to avoid repeating mistakes and to improve organizational and project processes and procedures. Lessons learned can be project oriented or product oriented. They can include information on risks, issues, procurements, quality defects, and any areas of poor or outstanding performance. A lessons learned register includes:

- Identifier
- Category
- Trigger
- Lesson
- Responsible party
- Comments

The lessons learned register is an output from process 4.4 Manage Project Knowledge in the *PMBOK®
Guide* – Sixth Edition. It is a dynamic document that is updated throughout the project.

Tailoring Tips

Consider the following tips to help tailor the lessons learned register to meet your needs:

- You can add information on the person identifying the lesson, especially if the person identifying the lesson and the person accountable for implementing it are different.
- Information on the next implementation opportunity and the expected implementation date can be used to ensure that the information isn't just recorded, but it is acted on as well.
- You can add a checkbox to indicate whether the lesson impacts an organizational system, policy or practice, or whether it can be implemented without the need to escalate up through the organization.

Alignment

The lessons learned register should be aligned and consistent with the following documents:

- Change management plan
- Change log
- Issue log
- Decision log
- Lessons learned summary

Description

You can use the element descriptions in the Table 3.5 to assist you in developing the lessons learned register.

TABLE 3.5 **Elements of a Lessons Learned Register**

Document Element	Description
ID	Enter a unique lesson identifier.
Category	Document the category of lesson, such as process, technical, environmental, stakeholder, phase, etc.
Trigger	Describe the context, event, or condition that led to the challenge, problem, or beneficial outcome.
Lesson	Articulate the lesson that can be passed on to other projects and to the organization.
Responsible party	Identify the person who is assigned to implement any changes to ensure the lesson is communicated and distributed.
Comments	Document any clarifying comments about the challenge, problem, good practice, or other fields on the form.

LESSONS LEARNED REGISTER

Project Title: _____

Date Prepared: _____

ID	Category	Trigger	Lesson	Responsible Party	Comments

3.6 QUALITY AUDIT

A quality audit is a technique that employs a structured, independent review to project and/or product elements. Any aspect of the project or product can be audited. Common areas for audit include:

- Project processes
- Project documents
- Product requirements
- Product documentation
- Defect or deficiency repair
- Compliance with organizational policies and procedures
- Compliance with the quality management plan
- Good practices from similar projects
- Areas for improvement
- Description of deficiencies or defects

Defects or deficiencies should include action items, a responsible party, and be assigned a due date for compliance.

A quality audit is a technique from process 8.2 Manage Quality in the *PMBOK® Guide* – Sixth Edition. Audits should be tailored to best meet the needs of the project.

Tailoring Tips

Consider the following tips to help tailor the quality audit to meet your needs:

- Quality audits can also include information that will be shared with other projects.
- Some projects use audits to track the implementation of approved changes and corrective or preventive actions.

Alignment

The quality audit should be aligned and consistent with the following documents:

- Quality management plan

Description

You can use the element descriptions in Table 3.6 to assist you in developing the quality audit.

TABLE 3.6   Elements of a Quality Audit

Document Element	Description		
Area audited	Check the box for the area or areas audited.		
Good practices from similar projects	Describe any good or best practices that can be shared from similar projects.		
Areas for improvement	Describe any areas that need improvement and the specific improvements or measurements that need to be achieved.		
Deficiencies or defects		ID	Enter a unique defect identifier.
		Defect	Describe the deficiency or defect.
		Action	Describe the corrective actions needed to fix the defect.
		Responsible party	Identify the person assigned to correct the deficiency or defect.
		Due date	Document the due date.
Comments	Provide any additional useful comments about the audit.		

QUALITY AUDIT

Project Title: _____ **Date Prepared:** _____

Project Auditor: _____ **Audit Date:** _____

Area Audited:

☐ Project processes	☐ Project documents
☐ Product documents	☐ Product documentation
☐ Quality management plan	☐ Defect/deficiency repair
☐ Organizational policies and procedures	

Good Practices from Similar Projects:

Areas for Improvement:

Deficiencies or Defects:

ID	Defect	Action	Responsible Party	Due Date

Comments:

3.7 TEAM PERFORMANCE ASSESSMENT

The team performance assessment is used to review technical and interpersonal competencies of the team as a whole, as well as general characteristics such as team morale and cohesiveness. It is used by the project manager to identify areas to improve the team's ability to achieve agreed-upon project objectives. The contents of the team performance assessment can include:

- Technical performance
  - Scope
  - Quality
  - Schedule
  - Cost
- Interpersonal competency
  - Communication
  - Collaboration
  - Conflict management
  - Decision making
- Team characteristics
  - Morale
  - Cohesiveness

The team performance assessment is an output of process 9.4 Develop Team in the *PMBOK® Guide* – Sixth Edition.

Tailoring Tips

Consider the following tips to help tailor the team performance assessment to meet your needs:

- For a small project you can just do a summary of technical performance and interpersonal competency rather than decomposing those categories into subcategories.
- If part of your role as a project manager is to evaluate individual team member performance you can modify the form to focus on individuals rather than the team.
- Some team assessments include team strengths and weaknesses.

Alignment

The team performance assessment should be aligned and consistent with the following documents:

- Resource management plan

Description

You can use the element descriptions in Table 3.7 to assist you in developing the team performance assessment.

TABLE 3.7 Elements of a Team Performance Assessment

Document Element		Description
Technical performance	Scope	Rate the team's ability to deliver the scope of the project and product. Provide comments that describe instances or aspects of scope performance that justify the rating.
	Quality	Rate the team's ability to deliver the quality required of the project and product. Provide comments that describe instances or aspects of quality performance that justify the rating.
	Schedule	Rate the team's ability to deliver on schedule. Provide comments that describe instances or aspects of schedule performance that justify the rating.
	Cost	Rate the team's ability to deliver within budget. Provide comments that describe instances or aspects of cost performance that justify the rating.
Interpersonal competency	Communication	Rate the team's ability to communicate effectively. Provide comments that illustrate instances of communication that justify the rating.
	Collaboration	Rate the team's ability to collaborate effectively. Provide comments that illustrate instances of collaboration that justify the rating.
	Conflict management	Rate the team's ability to manage conflict effectively. Provide comments that illustrate instances of conflict management that justify the rating.
	Decision making	Rate the team's ability to make decisions effectively. Provide comments that illustrate instances of decision making that justify the rating.
Team morale	Describe the overall team morale.	
Areas for development	Area	List technical or interpersonal areas for development.
	Approach	Describe the development approach, such as training, mentoring, or coaching.
	Actions	List the actions necessary to implement the development approach.

TEAM PERFORMANCE ASSESSMENT

Project Title: _____ **Date Prepared:** _____

Technical Performance

Scope	☐ Exceeds Expectations	☐ Meets Expectations	☐ Needs Improvement
Comments:			

Quality	☐ Exceeds Expectations	☐ Meets Expectations	☐ Needs Improvement
Comments:			

Schedule	☐ Exceeds Expectations	☐ Meets Expectations	☐ Needs Improvement
Comments:			

Cost	☐ Exceeds Expectations	☐ Meets Expectations	☐ Needs Improvement
Comments:			

TEAM PERFORMANCE ASSESSMENT

Interpersonal Competency

Communication	☐ Exceeds Expectations	☐ Meets Expectations	☐ Needs Improvement
Comments:			

Collaboration	☐ Exceeds Expectations	☐ Meets Expectations	☐ Needs Improvement
Comments:			

Conflict Management	☐ Exceeds Expectations	☐ Meets Expectations	☐ Needs Improvement
Comments:			

Decision Making	☐ Exceeds Expectations	☐ Meets Expectations	☐ Needs Improvement
Comments:			

TEAM PERFORMANCE ASSESSMENT

Team Morale	☐ Exceeds Expectations	☐ Meets Expectations	☐ Needs Improvement

Comments:

Areas for Development

Area	Approach	Actions

Monitoring and Controlling Forms

4.0 MONITORING AND CONTROLLING PROCESS GROUP

The purpose of the Monitoring and Controlling Process Group is to review project work results and compare them to planned results. A significant variance indicates the need for preventive actions, corrective actions, or change requests. There are 11 processes in the Monitoring and Controlling Process Group:

- Monitor and Control Project Work
- Perform Integrated Change Control
- Validate Scope
- Control Scope
- Control Schedule
- Control Costs
- Control Quality
- Control Resources
- Monitor Communications
- Monitor Risks
- Control Procurements
- Monitor Stakeholder Engagement

The intent of the Monitoring and Controlling Process Group is to at least:

- Review and analyze project performance
- Recommend changes and corrective and preventive actions
- Process change requests
- Report project performance
- Monitor risk activities, responses, and status
- Manage contractors
- Monitor the effectiveness of stakeholder engagement

Monitoring and controlling takes place throughout the project, from inception to closing. All variances are identified, and all change requests are processed here. The product deliverables are also accepted in the monitoring and controlling processes.

The forms used to document these activities include:

- Team member status report
- Project status report
- Variance analysis
- Earned value analysis
- Risk audit
- Contractor status report
- Procurement audit
- Contract closeout
- Product acceptance

4.1 TEAM MEMBER STATUS REPORT

The team member status report is filled out by team members and submitted to the project manager on a regular basis. It tracks schedule, quality, and cost status for the current reporting period and provides planned information for the next reporting period. Status reports also identify new risks and issues that have arisen in the current reporting period. Typical information includes:

- Activities planned for the current reporting period
- Activities accomplished in the current reporting period
- Activities planned but not accomplished in the current reporting period
- Root causes of activities variances
- Funds spent in the current reporting period
- Funds planned to be spent for the current reporting period
- Root causes of funds variances
- Quality variances identified in the current reporting period
- Planned corrective or preventive action
- Activities planned for the next reporting period
- Costs planned for the next reporting period
- New risks identified
- New issues identified
- Comments

This information is generally compiled by the project manager into a project status report. The team member status report and the project status report are examples of work performance reports, as mentioned in 4.5 Monitor and Control Project Work in the *PMBOK® Guide* – Sixth Edition. This report is submitted at predefined intervals throughout the project.

Tailoring Tips

Consider the following tips to help tailor the team member status report to meet your needs:

- You can add a field for escalations to identify those areas that need to be escalated to the sponsor, program manager, or other appropriate individual.
- Some reports include a field to record decisions made. These would be transferred to the project decision log.
- If your organization has a robust knowledge management process you might consider adding fields for knowledge transfer or lessons learned. These can then be transferred to the organization's knowledge repository or lessons learned register.

Alignment

The team member status report should be aligned and consistent with the following documents:

- Project schedule
- Cost estimates
- Project budget
- Issue log
- Risk register
- Project status report
- Variance analysis
- Earned value status report

Description

You can use the element descriptions in Table 4.1 to assist you in developing the team member status report.

TABLE 4.1   Elements of a Team Member Status Report

Document Element	Description
Activities planned this reporting period	List all activities scheduled for this period, including work to be started, continued, or completed.
Activities accomplished this reporting period	List all activities accomplished this period, including work that was started, continued, or completed.
Activities planned but not accomplished this reporting period	List all activities that were scheduled for this period, but not started, continued, or completed.
Root cause of variances	For any work that was not accomplished as scheduled, identify the cause of the variance.
Funds spent this reporting period	Record funds spent this period.
Funds planned to be spent this reporting period	Record funds that were planned to be spent this period.
Root cause of variances	For any expenditures that were over or under plan, identify the cause of the variance. Include information on labor vs. material variances. Identify if the basis of estimates or the assumptions were inaccurate.
Quality variances identified this period	Identify any product performance or quality variance.
Planned corrective of preventive action	Identify any actions needed to recover cost, schedule, or quality variances or prevent future variances.
Activities planned for next reporting period	List all activities scheduled for next period, including work to be started, continued, or completed.
Costs planned for next reporting period	Identify funds planned to be expended next period.
New risks identified	Identify any new risks that have arisen. New risks should be recorded in the risk register as well.
Issues	Identify any new issues that have arisen. New issues should be recorded in the issue log as well.
Comments	Document any comments that add relevance to this report.

TEAM MEMBER STATUS REPORT

Project Title: _____ **Date Prepared:** _____

Team Member: _____ **Role:** _____

Activities Planned for This Reporting Period

1.

2.

3.

4.

5.

6.

Activities Accomplished This Reporting Period

1.

2.

3.

4.

5.

6.

Activities Planned but Not Accomplished This Reporting Period

1.

2.

3.

4.

TEAM MEMBER STATUS REPORT

Root Cause of Activity Variances

Funds Spent This Reporting Period

Funds Planned to Be Spent This Reporting Period

Root Cause of Cost Variances

Quality Variances Identified This Period

Planned Corrective or Preventive Action

TEAM MEMBER STATUS REPORT

Activities Planned for Next Reporting Period

1.
2.
3.
4.
5.

Costs Planned for Next Reporting Period

New Risks Identified

Risk

New Issues Identified

Issue

Comments

4.2 PROJECT STATUS REPORT

The project status report (sometimes known as a performance report or progress report) is filled out by the project manager and submitted on a regular basis to the sponsor, project portfolio management group, project management office (PMO), or other project oversight person or group. The information is compiled from the team member status reports and includes overall project performance. It contains summary-level information, such as accomplishments, rather than detailed activity-level information. The project status report tracks schedule and cost status for the current reporting period and provides planned information for the next reporting period. It indicates impacts to milestones and cost reserves as well as identifying new risks and issues that have arisen in the current reporting period. Typical information includes:

- Accomplishments for the current reporting period
- Accomplishments planned but not completed in the current reporting period
- Root causes of accomplishment variances
- Impact to upcoming milestones or project due date
- Planned corrective or preventive action
- Funds spent in the current reporting period
- Root causes of budget variances
- Impact to overall budget or contingency funds
- Planned corrective or preventive action
- Accomplishments planned for the next reporting period
- Costs planned for the next reporting period
- New risks identified
- Issues
- Comments

The project status report is an example of a work performance report, an output of 4.5 Monitor and Control Project Work in the *PMBOK® Guide* – Sixth Edition. This report is submitted at predefined intervals throughout the project.

Tailoring Tips

Consider the following tips to help tailor the project status report to meet your needs:

- You can add a field for escalations to identify those areas that need to be escalated to the sponsor, program manager, or other appropriate individuals.
- Some reports include a field to record decisions made. These would be transferred to the project decision log.
- If there were any change requests that were submitted during the reporting period you may want to summarize them and refer the reader to the change log.
- If your organization has a robust knowledge management process you might consider adding fields for knowledge transfer or lessons learned. These can then be transferred to the organization's knowledge repository or lessons learned register.
- In addition to tailoring the content of the project status report, you can tailor the presentation. Many PMOs have reporting software that transforms the data into dashboards, heat reports, stop light charts, or other representations.

Alignment

The project status report should be aligned and consistent with the following documents:

- Team member status reports
- Project schedule

- Cost estimates
- Project budget
- Issue log
- Risk register
- Variance analysis
- Earned value status report
- Contractor status report

Description

You can use the element descriptions in Table 4.2 to assist you in developing the project status report.

TABLE 4.2   Elements of a Project Status Report

Document Element	Description
Accomplishments for this reporting period	List all work packages or other accomplishments scheduled for completion for the current reporting period.
Accomplishments planned but not completed this reporting period	List all work packages or other accomplishments scheduled for the current period but not completed.
Root cause of variances	Identify the cause of the variance for any work that was not accomplished as scheduled for the current period.
Impact to upcoming milestones or project due date	Identify any impact to any upcoming milestones or overall project schedule for any work that was not accomplished as scheduled. Identify any work currently behind on the critical path or if the critical path has changed based on the variance.
Planned corrective or preventive action	Identify any actions needed to make up schedule variances or prevent future schedule variances.
Funds spent this reporting period	Record funds spent this period.
Root cause of variance	Identify the cause of the variance for any expenditure over or under plan. Include information on the labor variance versus material variance and whether the variance is due to the basis of estimates or estimating assumptions.
Impact to overall budget or contingency funds	Indicate the impact to the overall project budget or whether contingency funds must be expended.
Planned corrective or preventive action	Identify any actions needed to recover cost variances or to prevent future schedule variances.
Accomplishments planned for next reporting period	List all work packages or accomplishments scheduled for completion next period.
Costs planned for next reporting period	Identify funds planned to be expended next period.
New risks identified	Identify any new risks that have been identified this period. These risks should be recorded in the risk register as well.
Issues	Identify any new issues that have arisen this period. These issues should be recorded in the issue log as well.
Comments	Record any comments that add relevance to the report.

PROJECT STATUS REPORT

Project Title: _____ Date Prepared: _____

Project Manager: _____ Sponsor: _____

Accomplishments for This Reporting Period

1.
2.
3.
4.
5.
6.

Accomplishments Planned but Not Completed This Reporting Period

1.
2.
3.
4.

Root Cause of Variances

PROJECT STATUS REPORT

Impact to Upcoming Milestones or Project Due Date

Planned Corrective or Preventive Action

Funds Spent This Reporting Period

Root Cause of Variances

PROJECT STATUS REPORT

Impact to Overall Budget or Contingency Funds

Planned Corrective or Preventive Action

Accomplishments Planned for Next Reporting Period

1.
2.
3.
4.

Costs Planned for Next Reporting Period

PROJECT STATUS REPORT

New Risks Identified

Risk:

Issues

Issue:

Comments

4.3 VARIANCE ANALYSIS

Variance analysis reports collect and assemble information on project performance variances. Common topics are schedule, cost, and quality variances. Information on a variance analysis includes:

- Schedule variance
  - Planned results
  - Actual results
  - Variance
  - Root cause
  - Planned response
- Cost variance
  - Planned results
  - Actual results
  - Variance
  - Root cause
  - Planned response
- Quality variance
  - Planned results
  - Actual results
  - Variance
  - Root cause
- Planned response

Variance analysis is an example of a data analysis technique in the *PMBOK® Guide* – Sixth Edition. The information can be provided as a standalone report, as part of the project status report, or as backup to an earned value status report. It is identified as a technique in these processes:

- 4.5 Monitor and Control Project Work
- 4.7 Close Project or Phase
- 5.6 Control Scope
- 6.6 Control Schedule
- 7.4 Control Costs

Tailoring Tips

Consider the following tips to help tailor the variance analysis to meet your needs:

- Scope variance can be included but is generally indicated by a schedule variance, as either more or less scope will have been accomplished over time.
- The variance analysis can be done at an activity, resource, work package, control account, or project level depending on your needs.
- You can add a check box to indicate if the information needs to be escalated to the sponsor, program manager, or other appropriate individuals.
- You may want to add a field that indicates the implications of continued variance. This can include a forecast based on a trend analysis or based on identified responses.
- In addition to tailoring the content of the variance analysis, you can tailor the presentation. Many PMOs have reporting software that transforms the data into dashboards, heat reports, stop light charts, or other representations.

Alignment

The variance analysis should be aligned and consistent with the following documents:

- Team member status reports
- Project status report
- Project schedule
- Cost estimates
- Project budget
- Issue log
- Earned value status report
- Contractor status report

Description

You can use the element descriptions in Table 4.3 to assist you in developing the variance analysis.

TABLE 4.3   Elements of Variance Analysis

Document Elements	Description	
Schedule variance	Planned result	Describe the work planned to be accomplished.
	Actual result	Describe the work actually accomplished.
	Variance	Describe the variance.
	Root cause	Identify the root cause of the variance.
	Planned response	Document the planned corrective or preventive action.
Cost variance	Planned result	Record the planned costs for the work planned to be accomplished.
	Actual result	Record the actual costs expended.
	Variance	Calculate the variance.
	Root cause	Identify the root cause of the variance.
	Planned response	Document the planned corrective or preventive action.
Quality variance	Planned result	Describe the planned performance or quality measurements.
	Actual result	Describe the actual performance or quality measurements.
	Variance	Describe the variance.
	Root cause	Identify the root cause of the variance.
	Planned response	Document the planned corrective action.

VARIANCE ANALYSIS

Project Title: _____ Date Prepared: _____

Schedule Variance

Planned Result	Actual Result	Variance

Root Cause

Planned Response

Cost Variance

Planned Result	Actual Result	Variance

VARIANCE ANALYSIS

Root Cause

Planned Response

Quality Variance

Planned Result	Actual Result	Variance

Root Cause

Planned Response

4.4 EARNED VALUE ANALYSIS

Earned value analysis shows specific mathematical metrics that are designed to reflect the health of the project by integrating scope, schedule, and cost information. Information can be reported for the current reporting period and on a cumulative basis. Earned value analysis can also be used to forecast the total cost of the project at completion or the efficiency required to complete the project for the baseline budget. Information that is generally collected includes:

- Budget at completion (BAC)
- Planned value (PV)
- Earned value (EV)
- Actual cost (AC)
- Schedule variance (SV)
- Cost variance (CV)
- Schedule performance index (SPI)
- Cost performance index (CPI)
- Percent planned
- Percent earned
- Percent spent
- Estimates at completion (EAC)
- To complete performance index (TCPI)

Earned value analysis is an example of a data analysis technique in the *PMBOK® Guide* – Sixth Edition. The information can be provided as a standalone report or as part of the project status report. Earned value analysis is conducted at pre-defined intervals throughout the project. It is identified as a technique in these processes:

- 4.5 Monitor and Control Project Work
- 6.6 Control Schedule
- 7.4 Control Costs
- 12.3 Control Procurements

Tailoring Tips

Consider the following tips to help tailor earned value analysis to meet your needs:

- The earned value analysis can be done at the control account and/or project level depending on your needs.
- You may want to add a field that indicates the implications of continued variance. This can include a forecast based on a trend analysis or based on identified responses.
- Several different equations can be used to calculate the EAC depending on whether the remaining work will be completed at the budgeted rate or at the current rate. Two options are presented on this form.
- There are options to calculate a TCPI. Use the information from your project to determine the best approach for reporting.
- You may want to add information that indicates the implications of continued schedule variance. This can include a schedule forecast using SPI as the basis for a trend analysis or based on analyzing the critical path.
- Some organizations are starting to embrace earned schedule metrics. You can update this form to include various earned schedule calculations in addition to a critical path analysis.
- In addition to tailoring the content of the earned value analysis, you can tailor the presentation. Many PMOs have reporting software that transforms the data into dashboards, control charts, S-curves, or other representations.

Alignment

Earned value analysis should be aligned and consistent with the following documents:

- Project status report
- Project schedule
- Project budget
- Variance analysis
- Contractor status report

Description

You can use the element descriptions in Table 4.4 to assist you in developing an earned value analysis.

TABLE 4.4　Elements of Earned Value Analysis

Document Element	Description
Planned value	Enter the value of the work planned to be accomplished.
Earned value	Enter the value of the work actually accomplished.
Actual cost	Enter the cost for the work accomplished.
Schedule variance	Calculate the schedule variance by subtracting the planned value from the earned value. $SV = EV - PV$
Cost variance	Calculate the cost variance by subtracting the actual cost from the earned value. $CV = EV - AC$
Schedule performance index	Calculate the schedule performance index by dividing earned value by the planned value. $SPI = EV/PV$
Cost performance index	Calculate the cost performance index by dividing the earned value by the actual cost. $CPI = EV/AC$
Root cause of schedule variance	Identify the root cause of the schedule variance.
Schedule impact	Describe the impact on deliverables, milestones, or critical path.
Root cause of cost variance	Identify the root cause of the cost variance.
Budget impact	Describe the impact on the project budget, contingency funds and reserves, and any intended actions to address the variance.
Percent planned	Indicate the cumulative percent of the work planned to be accomplished. PV/BAC
Percent earned	Indicate the cumulative percent of work that has been accomplished. EV/BAC
Percent spent	Indicate the total costs spent to accomplish the work. AC/BAC
Estimates at completion	Determine an appropriate method to forecast the total expenditures at the project completion. Calculate the forecast and justify the reason for selecting the particular estimate at completion. For example: If the CPI is expected to remain the same for the remainder of the project: $EAC = BAC/CPI$ If both the CPI and SPI will influence the remaining work: $EAC = AC + [(BAC - EV)/(CPI \times SPI)]$
To complete performance index	Calculate the work remaining divided by the funds remaining. $TCPI = (BAC - EV)/(BAC - AC)$ to complete on plan, or $TCPI = (BAC - EV)/(EAC - AC)$ to complete the current EAC.

EARNED VALUE ANALYSIS REPORT

Project Title: _____ Date Prepared: _____

Budget at Completion (BAC): _____ Overall Status: _____

	Current Reporting Period	Current Period Cumulative	Past Period Cumulative
Planned value (PV)			
Earned value (EV)			
Actual cost (AC)			
Schedule variance (SV)			
Cost variance (CV)			
Schedule performance index (SPI)			
Cost performance index (CPI)			
Root Cause of Schedule Variance:			
Schedule Impact:			
Root Cause of Cost Variance:			
Budget Impact:			
Percent planned			
Percent earned			
Percent spent			
Estimates at Completion (EAC):			
EAC w/CPI [BAC/CPI]			
EAC w/ CPI*SPI [AC + ((BAC - EV)/ (CPI*SPI))]			
Selected EAC, Justification, and Explanation			
To complete performance index (TCPI)			

4.5 RISK AUDIT

Risk audits are used to evaluate the effectiveness of the risk identification, risk responses, and risk management process as a whole. Information reviewed in a risk audit can include:

- Risk event audits
  - Risk events
  - Causes
  - Responses
- Risk response audits
  - Risk event
  - Responses
  - Success
  - Actions for improvement
- Risk management processes
  - Process
  - Compliance
  - Tools and techniques used
- Good practices
- Areas for improvement

The risk audit is a tool used in process 11.7 Control Risks in the *PMBOK® Guide* – Sixth Edition. It is conducted periodically as needed.

Tailoring Tips

Consider the following tips to help tailor the risk audit to meet your needs:

- To make the audit more robust you can include an assessment of the effectiveness of the risk management approach.
- Large organizations often have policies and procedures associated with project risk management. If this is the case in your organization, include an assessment of compliance with the policies and procedures.
- Many organizations don't track opportunity management. You can expand the scope of the audit to include opportunity management if appropriate.
- For larger projects you may want to include information on overall risk in addition to risk events.

Alignment

The risk audit should be aligned and consistent with the following documents:

- Risk management plan
- Risk register
- Risk report

Description

You can use the element descriptions in Table 4.5 to assist you in developing the risk audit.

TABLE 4.5   Elements of Risk Audit

Document Element	Description	
Risk event audit	Event	List the event from the risk register.
	Cause	Identify the root cause of the event from the risk register.
	Response	Describe the response implemented.
	Comment	Discuss if there was any way to have foreseen the event and respond to it more effectively.
Risk response audit	Event	List the event from the risk register.
	Response	List the risk response from the risk register.
	Successful	Indicate if the response was successful.
	Actions to improve	Identify any opportunities for improvement in risk response.
Risk management process audit	Plan risk management	Followed: Indicate if the various processes were followed as indicated in the risk management plan.
	Identify risks	Tools and techniques used: Identify tools and techniques used in the various risk management processes and whether they were successful.
	Perform qualitative risk analysis	
	Perform quantitative risk analysis	
	Plan risk responses	
	Control risks	
Description of good practices to share	Describe any practices that should be shared for use on other projects. Include any recommendations to update and improve risk forms, templates, policies, procedures, or processes to ensure these practices are repeatable.	
Description of areas for improvement	Describe any practices that need improvement, the improvement plan, and any follow-up dates or information for corrective action.	

RISK AUDIT

Project Title: _____ Date Prepared: _____

Project Auditor: _____ Audit Date: _____

Risk Event Audit

Event	Cause	Response	Comment

Risk Response Audit

Event	Response	Successful	Actions to Improve

Risk Management Process Audit

Process	Followed	Tools and Techniques Used
Plan Risk Management		
Identify Risks		
Perform Qualitative Assessment		
Perform Quantitative Assessment		
Plan Risk Responses		
Monitor and Control Risks		

RISK AUDIT

Description of Good Practices to Share

Description of Areas for Improvement

4.6 CONTRACTOR STATUS REPORT

The contractor status report is filled out by the contractor and submitted on a regular basis to the project manager. It tracks status for the current reporting period and provides forecasts for future reporting periods. The report also gathers information on new risks, disputes, and issues. Information can include:

- Scope performance
- Quality performance
- Schedule performance
- Cost performance
- Forecasted performance
- Claims or disputes
- Risks
- Preventive or corrective action
- Issues

This information is generally included in the project status report compiled by the project manager. The contractor status report is an example of work performance information identified in 12.3 Control Procurements in the *PMBOK® Guide* – Sixth Edition. This report is submitted at pre-defined intervals throughout the project.

Tailoring Tips

Consider the following tips to help tailor the contractor status report to meet your needs:

- You can add a field for escalations to identify those areas that need to be escalated to the sponsor, program manager, contracting officer, or other appropriate individuals.
- If there were any contract change requests that were submitted during the reporting period, summary information should be described in the contractor status report.
- In addition to tailoring the content of the contractor status report, you can tailor the presentation. Many PMOs have reporting software that transforms the data into dashboards, heat reports, stop light charts, or other representations.

Alignment

The contractor status report should be aligned and consistent with the following documents:

- Procurement management plan
- Project schedule
- Cost estimates
- Project budget
- Variance analysis
- Earned value status report
- Project status report

Description

You can use the element descriptions in Table 4.6 to assist you in developing the contractor status report.

TABLE 4.6   Elements of a Contractor Status Report

Document Element	Description
Scope performance this reporting period	Describe progress on scope made during this reporting period.
Quality performance this reporting period	Identify any quality or performance variances.
Schedule performance this reporting period	Describe whether the contract is on schedule. If ahead or behind, identify the cause of the variance.
Cost performance this reporting period	Describe whether the contract is on budget. If over or under budget, identify the cause of the variance.
Forecast performance for future reporting periods	Discuss the estimated delivery date and final cost of the contract. If the contract is a fixed price, do not enter cost forecasts.
Claims or disputes	Identify any new or resolved disputes or claims that have occurred during the current reporting period.
Risks	List any risks. Risks should also be in the risk register.
Planned corrective or preventive action	Identify planned corrective or preventive actions necessary to recover schedule, cost, scope, or quality variances.
Issues	Identify any new issues that have arisen. These should also be entered in the issue log.
Comments	Add any comments that will add relevance to the report.

CONTRACTOR STATUS REPORT

Project Title: _____ Date Prepared: _____

Vendor: _____ Contract #: _____

Scope Performance This Reporting Period

[ ]

Quality Performance This Reporting Period

[ ]

Schedule Performance This Reporting Period

[ ]

CONTRACTOR STATUS REPORT

Cost Performance This Reporting Period

Forecast Performance for Future Reporting Periods

Claims or Disputes

Risks

CONTRACTOR STATUS REPORT

Planned Corrective or Preventive Action

Issues

Comments

4.7 PROCUREMENT AUDIT

The procurement audit is the review of contracts and contracting processes for completeness, accuracy, and effectiveness. Information in the audit can be used to improve the process and results on the current procurement or on other contracts. Information recorded in the audit includes:

- Vendor performance audit
  - Scope
  - Quality
  - Schedule
  - Cost
  - Other information
- Procurement management process audit
  - Process
  - Tools and techniques used
- Description of good practices
- Description of areas for improvement

The procurement audit is a technique from process 12.3 Control Procurements in the *PMBOK® Guide –* Sixth Edition. It is conducted periodically throughout the project, or as needed.

Tailoring Tips

Consider the following tips to help tailor the procurement audit to meet your needs:

- Add qualitative information, such as how easy the vendor was to work with, timeliness of returning calls, and collaborative attitude. This can provide useful information for future procurement opportunities.

Alignment

The procurement audit should be aligned and consistent with the following documents:

- Procurement management plan
- Contractor status report
- Contract closeout report

Description

You can use the element descriptions in Table 4.7 to assist you in developing the procurement audit.

TABLE 4.7   Elements of a Procurement Audit

Document Element	Description		
What worked well	Scope	Describe aspects of contract scope that were handled well.	
	Quality	Describe aspects of product quality that were handled well.	
	Schedule	Describe aspects of the contract schedule that were handled well.	
	Cost	Describe aspects of the contract budget that were handled well.	
	Other	Describe any other aspects of the contract or procurement that were handled well.	
What can be improved	Scope	Describe aspects of contract scope that could be improved.	
	Quality	Describe aspects of product quality that could be improved.	
	Schedule	Describe aspects of the contract schedule that could be improved.	
	Cost	Describe aspects of the contract budget that could be improved.	
	Other	Describe any other aspects of the contract or procurement that could be improved.	
Procurement management process audit	Plan procurements	Indicate if each procurement was followed or not.	Describe any tools or techniques that were effective for each procurement.
	Conduct procurements		
	Control procurements		
Good practices to share	Describe any good practices that can be shared with other projects or that should be incorporated into organization policies, procedures, or processes. Include information on lessons learned.		
Areas for improvement	Describe any areas that should be improved with the procurement process. Include information that should be incorporated into policies, procedures, or processes. Include information on lessons learned.		

PROCUREMENT AUDIT

Project Title: _____ Date Prepared: _____

Project Auditor: _____ Audit Date: _____

Vendor Performance Audit

What Worked Well:	
Scope	
Quality	
Schedule	
Cost	
Other	
What Can Be Improved:	
Scope	
Quality	
Schedule	
Cost	
Other	

Procurement Management Process Audit

Process	Followed	Tools and Techniques Used
Plan Procurements		
Conduct Procurements		
Control Procurements		

PROCUREMENT AUDIT

Description of Good Practices to Share

Description of Areas for Improvement

4.8 CONTRACT CLOSEOUT REPORT

Contract closeout involves documenting the vendor performance so that the information can be used to evaluate the vendor for future work. Contract closure supports the project closure process and helps ensure contractual agreements are completed or terminated. Before a contract can be fully closed or terminated, all disputes must be resolved, the product or result must be accepted, and the final payments must be made. Information recorded as part of closing out a contract includes:

- Vendor performance analysis
  - Scope
  - Quality
  - Schedule
  - Cost
  - Other information, such as how easy the vendor was to work with
- Record of contract changes
  - Change ID
  - Description of change
  - Date approved
- Record of contract disputes
  - Description of dispute
  - Resolution
  - Date resolved

The date of contract completion, who signed off on it, and the date of the final payment are other elements that should be recorded.

Tailoring Tips

Consider the following tips to help tailor the contract closeout report to meet your needs:

- For a small contract you can combine all the information in a vendor performance analysis into a summary paragraph.
- For small contracts you may not need information on contract changes or contract disputes.
- If the project was based around one large contract, you can combine the information in the project closeout report with this form.

Alignment

The contract closeout report should be aligned and consistent with the following documents:

- Procurement management plan
- Procurement audit
- Change log
- Project closeout

Description

You can use the element descriptions in Table 4.8 to assist you in developing the contract closeout report.

TABLE 4.8   Elements of a Contract Closeout

Document Element	Description	
What worked well	Scope	Describe aspects of contract scope that were handled well.
	Quality	Describe aspects of product quality that were handled well.
	Schedule	Describe aspects of the contract schedule that were handled well.
	Cost	Describe aspects of the contract budget that were handled well.
	Other	Describe any other aspects of the contract or procurement that were handled well.
What can be improved	Scope	Describe aspects of contract scope that could be improved.
	Quality	Describe aspects of product quality that could be improved.
	Schedule	Describe aspects of the contract schedule that could be improved.
	Cost	Describe aspects of the contract budget that could be improved.
	Other	Describe any other aspects of the contract or procurement that could be improved.
Record of contract changes	Change ID	Enter the change identifier from the change log.
	Change description	Enter the description from the change log.
	Date approved	Enter the date approved from the change log.
Record of contract disputes	Description	Describe the dispute or claim.
	Resolution	Describe the resolution.
	Date resolved	Enter the date the dispute or claim was resolved.

CONTRACT CLOSEOUT

Project Title: _____   Date Prepared: _____

Project Manager: _____   Contract Representative: _____

Vendor Performance Analysis

What Worked Well:	
Scope	
Quality	
Schedule	
Cost	
Other	
What Can Be Improved:	
Scope	
Quality	
Schedule	
Cost	
Other	

Record of Contract Changes

Change ID	Change Description	Date Approved

CONTRACT CLOSEOUT

Record of Contract Disputes

Description	Resolution	Date Resolved

Date of Contract Completion _____

Signed Off by _____

Date of Final Payment _____

4.9 PRODUCT ACCEPTANCE FORM

Product acceptance should be done periodically throughout the project as each deliverable or component is validated and accepted.

The product acceptance form can include this information:

- Identifier
- Requirement
- Acceptance criteria
- Method of validation
- Status
- Sign-off

Product acceptance is part of process 5.5 Validate Scope in the *PMBOK® Guide* – Sixth Edition. Product acceptance should be done periodically throughout the project life cycle.

Tailoring Tips

Consider the following tips to help tailor the product acceptance form to meet your needs:

- For a small project with only a few deliverables you may not need this form.
- You can add a column to indicate the verification method used to prove that the deliverables meet the requirements. In this instance it becomes a product verification, validation, and acceptance form.

Alignment

The product acceptance form should be aligned and consistent with the following documents:

- Scope management plan
- Requirements documentation
- Requirements traceability matrix
- Quality management plan

Description

You can use the element descriptions in Table 4.9 to assist you in developing the product acceptance form.

TABLE 4.9   Elements of Product Acceptance

Document Element	Description
ID	Enter a unique requirement identifier from the requirements documentation.
Requirement	Enter the requirement description from the requirements documentation.
Acceptance criteria	Enter the criteria for acceptance.
Validation method	Describe the method of validating the requirement meets the stakeholder's needs.
Status	Document whether the requirement or deliverable was accepted or not.
Sign-off	Obtain the signature of the party accepting the product.

FORMAL ACCEPTANCE FORM

Project Title: _____   Date Prepared: _____

ID	Requirement	Acceptance Criteria	Validation Method	Status	Comments	Sign-off

Closing

5.0 CLOSING PROCESS GROUP

The purpose of the Closing Process Group is to complete the project work, product work, and project phases in an orderly manner. There is one process in the Closing Process Group:

- Close Project or Phase

The intent of the Closing Process Group is to at least:

- Close project phases
- Close the project
- Document lessons learned
- Document final project results
- Archive project records
- Ensure the work necessary to close contracts is complete

As the final process in a phase or project, this process ensures an organized and efficient completion of deliverables, phases, and the project as a whole.

The forms used to document project closure include:

- Lessons learned
- Project closeout

5.1 LESSONS LEARNED SUMMARY

Lessons learned are compiled throughout the project or at specific intervals, such as at the end of a life cycle phase. These are recorded in the lessons learned register. The lessons learned summary compiles and organizes those things that the project team did that worked very well and should be passed along to other project teams, and identifies those things that should be improved for future project work. The summary should include information on risks, issues, procurements, quality defects, and any areas of poor or outstanding performance. Information that can be documented includes:

- Project performance analysis
  - Requirements
  - Scope
  - Schedule
  - Cost
  - Quality
  - Physical resources
  - Team resources
  - Communication
  - Reporting
  - Risk management
  - Procurement management
  - Stakeholder management
  - Process improvement
  - Product-specific information
- Information on specific risks
- Quality defects
- Vendor management
- Areas of exceptional performance
- Areas for improvement

This information is saved, along with the lessons learned register, in a lessons learned repository. Repositories can be as simple as a lessons learned binder, or they can be a searchable database, or anything in between. The purpose is to improve performance on the current project (if done during the project) and future projects. Use the information from your project to tailor the form to best meet your needs. The lessons learned summary supports process 4.7 Close Project or Phase in the *PMBOK® Guide* – Sixth Edition. It is developed at the close of a phase for long projects and at the close of the project for shorter projects.

Tailoring Tips

Consider the following tips to help tailor the lessons learned summary to meet your needs:

- Add, combine, or eliminate rows as needed to capture the important aspects of your project.
- Consider including a section on change management, as that can sometimes be a challenging aspect of projects to manage.
- You may want to include a section on phase management if you are doing the summary at the end of the project.
- Integration management is an important aspect for projects that are part of a program or for projects that have multiple organizations working on one project.
- If your project used a new development approach, such as a blended predictive and adaptive (agile) approach, consider adding some relevant lessons learned.

Alignment

The lessons learned summary should be aligned and consistent with the following documents:

- Issue register
- Risk register
- Decision log
- Lessons learned register
- Retrospectives

Description

You can use the element descriptions in Table 5.1 to assist you in developing the lessons learned summary.

TABLE 5.1   Elements of a Lessons Learned Summary

Document Element	Description	
Project Performance	What Worked Well	What Can Be Improved
Requirements definition and management	List any practices or incidents that were effective in defining and managing requirements.	List any practices or incidents that can be improved in defining and managing requirements.
Scope definition and management	List any practices or incidents that were effective in defining and managing scope.	List any practices or incidents that can be improved in defining and managing scope.
Schedule development and control	List any practices or incidents that were effective in developing and controlling the schedule.	List any practices or incidents that can be improved in developing and controlling the schedule.
Cost estimating and control	List any practices or incidents that were effective in developing estimates and controlling costs.	List any practices or incidents that can be improved in developing estimates and controlling costs.
Quality planning and control	List any practices or incidents that were effective in planning, managing, and controlling quality.	List any practices or incidents that can be improved in planning, managing, and controlling quality. Specific defects are addressed elsewhere.
Physical resource planning and control	List any practices or incidents that were effective in planning, acquiring, and managing physical resources.	List any practices or incidents that can be improved in planning, acquiring, and managing physical resources.
Team planning, development, and performance	List any practices or incidents that were effective in working with team members and developing and managing the team.	List any practices or incidents that can be improved in working with team members and developing and managing the team.
Communications management	List any practices or incidents that were effective in planning and distributing information.	List any practices or incidents that can be improved in planning and distributing information.
Reporting	List any practices or incidents that were effective in reporting project performance.	List any practices or incidents that can be improved in reporting project performance.
Risk management	List any practices or incidents that were effective in the risk management process. Specific risks are addressed elsewhere.	List any practices or incidents that can be improved in the risk management process. Specific risks are addressed elsewhere.

(continued)

TABLE 5.1   **Elements of a Lessons Learned Summary** *(continued)*

Document Element		Description
Project Performance	**What Worked Well**	**What Can Be Improved**
Procurement planning and management	List any practices or incidents that were effective in planning, conducting, and administering contracts.	List any practices or incidents that can be improved in planning, conducting, and administering contracts.
Stakeholder engagement	List any practices or incidents that were effective in engaging stakeholders.	List any practices or incidents that can be improved in engaging stakeholders.
Process improvement information	List any processes that were developed that should be continued.	List any processes that should be changed or discontinued.
Product-specific information	List any practices or incidents that were effective in delivering the specific product, service, or result.	List any practices or incidents that can be improved in delivering the specific product, service, or result.
Other	List any other practices or incidents that were effective, such as change control, configuration management, etc.	List any other practices or incidents that can be improved, such as change control, configuration management, etc.
Risks and issues	Risk or issue description	Identify risks or issues that occurred that should be considered to improve organizational learning.
	Response	Describe the response and its effectiveness.
	Comments	Provide any additional information needed to improve future project performance.
Quality defects	Defect description	Describe quality defects that should be considered to improve organizational effectiveness.
	Resolution	Describe how the defects were resolved.
	Comments	Indicate what should be done to improve future project performance.
Vendor management	Vendor	List the vendor(s).
	Issue	Describe any issues, claims, or disputes that occurred.
	Resolution	Describe the outcome or resolution.
	Comments	Indicate what should be done to improve future vendor management performance.
Other	Areas of exceptional performance	Identify areas of exceptional performance that can be passed on to other teams.
	Areas for improvement	Identify areas that can be improved on for future performance.

LESSONS LEARNED SUMMARY

Project Title: _____ **Date Prepared:** _____

Project Performance Analysis

	What Worked Well	What Can Be Improved
Requirements definition and management		
Scope definition and management		
Schedule development and control		
Cost estimating and control		
Quality planning and control		
Physical resource planning and control		
Team planning, development, and performance		
Communications management		
Reporting		
Risk management		
Procurement planning and management		
Stakeholder engagement		
Process improvement information		
Product-specific information		
Other		

LESSONS LEARNED SUMMARY

Risks and Issues

Risk or Issue Description	Response	Comments

Quality Defects

Defect Description	Resolution	Comments

Vendor Management

Vendor	Issue	Resolution	Comments

Other

Areas of Exceptional Performance	Areas for Improvement

5.2 PROJECT OR PHASE CLOSEOUT

Project closeout involves documenting the final project performance as compared to the project objectives. The objectives from the project charter are reviewed and evidence of meeting them is documented. If an objective was not met, or if there is a variance, that is documented as well. In addition, information from the procurement closeout is documented. Information documented includes:

- Project or phase description
- Project or phase objectives
- Completion criteria
- How met
- Cost and schedule variances
- Benefits management
- Business needs
- Summary of risks and issues

Use the information from your project to determine the best approach.

Tailoring Tips

Consider the following tips to help tailor the project or phase closeout to meet your needs:

- When working with an incremental life cycle or agile development method, the delivery of a major end item, service, or capability may benefit from a formal phase closeout report.
- For projects that are part of a program you should tailor the content to meet the needs of the program.

Alignment

The project or phase closeout should be aligned and consistent with the following documents:

- Project management plan (all components)
- Product acceptance
- Lessons learned summary

The project closeout form supports process 4.7 Close Project or Phase in the *PMBOK® Guide* – Sixth Edition.

Description

You can use the element descriptions in Table 5.2 to assist you in developing the project or phase closeout.

TABLE 5.2   Elements of a Project or Phase Closeout

Document Element	Description		
Project description	Provide a summary level description of the project.		
Performance summary	Scope		Describe the scope objectives needed to achieve the planned benefits of the project.
			Document the specific and measurable criteria needed to complete the scope objectives.
			Provide evidence that the completion criteria were met.
	Quality		Describe the quality objectives and criteria needed to achieve the planned benefits of the project.
			Document the specific and measurable criteria needed to meet the product and project quality objectives.
			Enter the verification and validation information from the product acceptance form.
Variances	Document the time and cost objectives and the final completion date and final expenditures. Explain any variances.		
Benefits management	Describe how the final product, service, or result achieved the benefits the project was undertaken to address.		
Business needs	Describe how the final product, service, or result achieved the business needs identified in the business plan.		
Risks and issues	Summarize any significant risks or issues, or the overall risk exposure, and describe the response and resolution strategies.		

PROJECT CLOSEOUT

Project Title: _____

Date Prepared: _____

Project Manager: _____

Project Description

PROJECT CLOSEOUT

Performance Summary

	Objectives	Completion Criteria	How Met
Scope			
Quality			

Variances

	Objectives/Final Outcome	Variances	Comments
Time			
Cost			

Benefits Management

Business Needs

Risks and Issues

Risk or Issue	Response or Resolution	Comments

Agile

New to this edition of *A Project Manager's Book of Forms* is a section with a few forms that can be used to apply agile techniques to your project. Many projects that use an agile development method have software that organizes and drives the work. Another method of organizing and tracking work in agile projects is by creating visual displays such as Kanban Boards to enable everyone to see the status of work in progress. Thus, there are not that many forms that are produced in traditional office software.

Some of the techniques used in agile development methods can be easily translated into a traditional form. The following forms are included and can be tailored for use in any project, regardless of the development method used:

- Project vision
- Backlog
- Release plan
- Retrospective

The *PMBOK® Guide* – Sixth Edition does not list these forms as outputs of any processes. The following table indicates the processes where they might be most useful.

Project Vision	Develop project charter.
	This form may be used prior to developing a charter, or, for smaller projects, instead of a charter.
Backlog	Collect requirements.
	The backlog can be used in place of or in addition to the requirements documentation.
Release Plan	Plan schedule management.
	The release plan is a high-level schedule. It may take the place of a schedule management plan or a high-level schedule for agile teams on small projects.
Retrospective	Manage project knowledge.
	The retrospective can take the place of or work in parallel with the lessons learned register.

6.1 PRODUCT VISION

The product vision provides the future view of the product being developed. The product vision is aspirational, yet achievable and realistic. It is developed at the very beginning of a project. It is often an input to the business case.

The product vision includes at least:

- Target customer
- Needs
- Product and key attributes
- Key benefits

The product vision is often used in place of a project charter on agile-based projects. It is developed once, at the beginning of the project.

Tailoring Tips

Consider the following tips to help tailor the product vision to meet your needs:

- Document the business goals that the product is aligned to.
- Identify key competitors and how this product will be better.
- Describe what differentiates this product from similar products in the market.
- You can combine all the information into a sentence or two, or you can follow a formula like the one shown in the sample form.

Alignment

The product vision should be aligned and consistent with the following documents:

- Product backlog
- Roadmap
- Release plan

Description

You can use the element descriptions in Table 6.1 to assist you in developing the product vision.

TABLE 6.1 Elements of a Product Vision

Document Element	Description
Target customer	The person or group who will buy or use the product.
Needs	The needs or requirements that the product will address.
Product and key attributes	A description of the product that includes attributes, functional and nonfunctional requirements, and top-level requirements.
Key benefit	Describes why the customer would buy the product.

PRODUCT VISION

Project Title: _____ **Date Prepared:** _____

We are developing this product for _____.

To respond to the following need(s):

-
-
-

This product responds to those needs by providing the following:

-
-
-
-

Customers will buy this product because:

-
-
-
-

6.2 PRODUCT BACKLOG

The product backlog is developed at the very beginning of a project. It is often developed in conjunction with the product vision. The product backlog keeps track of all the requirements along with their priority and the release they will be incorporated into.

The product backlog includes at least:

- ID
- Summary description
- Priority
- Story
- Status

The product backlog is used to document and prioritize the requirements, features, functions, and user stories for releases or sprints. It is developed at the start of the project and is updated throughout the project.

Tailoring Tips

Consider the following tips to help tailor the product backlog to meet your needs:

- You can add estimating information such as story points.
- To provide more detail you can indicate which sprint a feature or function will be incorporated into.
- You may want to indicate the user type that will benefit from the requirement, such as customer, administrator, manager, etc.
- For large projects it helps to categorize requirements, so having a column that indicates the category can be useful.

Alignment

The backlog should be aligned and consistent with the following documents:

- Product vision
- Roadmap
- Release plan

Description

You can use the element descriptions in Table 6.2 to assist you in developing the product backlog.

TABLE 6.2 Elements of a Product Backlog

Document Element	Description
ID	A unique identifier
Summary description	A brief description of the requirement or need. The description should be no more than one or two sentences.
Priority	A way of prioritizing or ranking the requirements. This can be in summary groups, such as high, medium and low, or it can be numbered 1, 2, 3.
Story	This field can either be a user story that is prioritized, or it can indicate the name of a user story that is recorded elsewhere.
Status	Indicates if the requirement is not started, in progress, or complete.

PRODUCT BACKLOG

Project Title: _____ Date Prepared: _____

ID	Summary Description	Priority	Story	Status

6.3 RELEASE PLAN

The release plan is similar to a roadmap. It functions as a high-level schedule that indicates which release each requirement or user story will be assigned to. The elements in a particular release can be updated based on the relative priority of the requirements in the backlog and the availability of resources needed to work on the specific requirements.

The release plan includes at least:

- Release dates
- User stories or requirements from the backlog

The release plan can be further elaborated into sprints. Each release may have multiple sprints. Once a sprint starts, the requirements or user stories in the sprint cannot be changed. A high-level release plan may be developed after the product vision and backlog are started; however, it will remain somewhat dynamic throughout the project as priorities shift and new requirements are identified. The sequence and priorities of requirements or user stories may be updated after each sprint to reflect changing needs based on performance feedback or customer needs.

Tailoring Tips

Consider the following tips to help tailor the release plan to meet your needs:

- As information in the release plan gets more concrete you may want to assign requirements or user stories to specific sprints in the release.
- You can arrange the release plan in swim lanes, with each lane assigned to a specific team or work stream.
- If you show the release plan with a timeline you can show the relationships between various user stories or features. This blends the information about the content of each release with the schedule information.

Alignment

The release plan should be aligned and consistent with the following documents:

- Product vision
- Roadmap
- Product backlog

Description

You can use the element descriptions in Table 6.3 to assist you in developing the release plan.

TABLE 6.3   Elements of a Release Plan

Document Element	Description
Release dates	Either a timeline or a milestone indicator of when releases start and finish. This can be more detailed to show a linear schedule that indicates the duration of each release.
User stories	The requirements or user stories from the backlog

RELEASE PLAN

Project Name: _____

Release Goal: Describe the goal of the release in this space.

This diagram assumes that different shades notes indicate different categories of user stories.

6.4 RETROSPECTIVE

The retrospective is an activity that is performed at the end of every sprint. The information is usually recorded on sticky notes or recorded in software. A common retrospective approach is called a "starfish." The starfish retrospective collects the following information:

- Start
- Stop
- Keep
- More
- Less

The intent of a retrospective is to improve the performance of the team and make them more efficient in each subsequent sprint.

Tailoring Tips

Consider the following tips to help tailor the retrospective to meet your needs:

- Instead of a starfish approach you can use "FLAP," which stands for Future Considerations, Lessons, Accomplishments, and Problems.
- You can color code information to indicate a category, such as technical, process, people, environment, etc.

Alignment

The retrospective should be aligned and consistent with the following documents:

- Lessons learned summary
- Project closeout

Description

You can use the element descriptions in Table 6.4 to assist you in developing the retrospective.

TABLE 6.4   Elements of a Retrospective

Document Element	Description
Start	Actions and behaviors that the team will begin to implement
Stop	Actions or behaviors that the team will cease doing
Keep	Practices that the team should continue with
More	Practices that were not done consistently that should be done more often
Less	Practices that were done too much or that should be reduced

RETROSPECTIVE

Project Title: _____   **Date Prepared:** _____

Start	Stop	Keep	More	Less

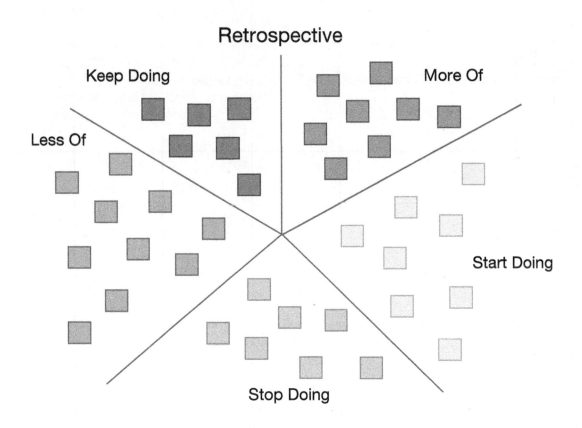

Index

Note: Page numbers followed by t denote tables; page numbers in *italics* denote the blank copy of the form.

A

Acceptance forms. *See* Product acceptance form; Product verification, validation, and acceptance form
Activity attributes, 62–63, 63t, *64*
Activity cost estimates, 82, 86t, *87*. *See also* Cost estimates
Activity duration estimates. *See* Duration estimates
Activity management
    Activity list, 59–60, 60t, *61*
    attributes, 62–63, 63t, *64*
    durations, 70, 74
    resources, 113, 116
Activity resource requirements. *See* Resource requirements
Agile development method, 230–248
    overview, viii, 239, 239t
    Product backlog, 242, 242t, *243*
    Product vision, 240, 240t, *241*
    Release plan, 244, 244t, *245*
    Retrospective, 239t, 246, 246t, *247–248*
Analogous estimates, 70, 73, 75t, *77,*85, 88, 89t, *91*
Assumption log, 9–10, 10t, *11,*63, 148, *151*
Authorization of projects, 3–4t, *4–8,*45, 235, 240

B

Backlogs for products, 59, 63, 239t, 242, 242t, *243*
Baselines
    for cost, 20, 22t, *24,*93, *94*
    for performance measurement, 20, 82, 83t, *84,*93
    for schedules, 20, 22t, *24*
    for scope, 20, 22t, *24,*37, 49, 52–53
Beta distribution/Beta distribution equation (three-point estimate), 70, 73, 75–76t, *77,* 85, 88, 89t, *91*
Bottom-up estimates, 88, 90t, *92*
Breakdown of project work
    resources, 116, *117*
    Work breakdown structure (WBS), 49–53, 50t, *51,*52, 54t, *55*
Budget at completion. *See* Earned value analysis
Bug fix documentation, 175

C

Change log, 175, 176t, *177*
Change management plan, 25–26, 26t, *27*
Change requests, 170, 171t, *172–174,*175, 196, 213
Change audit, 181
Charter, 2–3, 3–4t, *5–8,*45, 235, 240
Closing process, 229–238
    Closing process group, 229
    Lessons learned summary, 229–231, 231–232t, *233–234*
    Project (phase) closeout, 222, 235, 236t, *237–238*
    purpose, 229
Communications management plan, 118–119, 119t, *121*
Competency review, 184, 185t
Component relationships, 67–68, *69*
Configuration management, 25, 175
Constraints, 9–10, 10t, ,62, 148, *151*
Contingency reserve, 70, 85, 93
Contract change requests, 213
Contract closeout report, 222, 223t, *224–225*
Contractor status report, 213, 214t, *216–217*
Contracts and contracting process
    changes, 213
    closing report, 222, 223t, *224–225*
    disputes, 222, 223t, *225*
    procurement audits, 218, 219t, *220–221*
    project charter and, 2–3, 3–4t, *5–8,*45, 235, 240
    resource management, 82, 104–105, 106, *109–110*
    status report, 213, 214t, *216–217*
Control accounts, 50t, *51,*52
Control forms. *See* Monitoring and control forms
Control quality phase, 99
Cost estimates
    Activity cost estimates, 82, 86t, *87*
    Cost estimating worksheet, 88, 89t, 90t, *91–92*
    Cost estimation equation, 88, 89t
    overview, 85–86

Cost management
    baselines, 20, 22t, *24,*93, *94*
    Cost management plan, 82–83, 83t, *84*
    cumulative cost curve, 93
    earned value analysis, 56, 82, 93, 206–207,
        207t, *208*
    estimations of, 85–88, 86t, *87,*89–90t,
        91–92
    plan for, 82–83, 83t, *84*
    status report, 121, 191, 196–197, 198t,
        *199–201,*213
    variance analysis, 202–203, 203t,
        *204–205,*206–207, 207t, *208*
Cost performance index (CPI), 207t, *208*
Cumulative cost curve, 93

D
Decision log, 9, 168, 168t, *169,*196
Disputes, contract, 222, 223t, *225*
Documentation. *See* Project documentation
Duration estimates
    Duration estimating worksheet, 73–74,
        75–76t, *77*
    duration estimation equation, 73, 75t
    overview, 70–71, 71t, *72*

E
Earned value analysis, 206–207, 207t, *208*
Earned value management (EVM), 56, 82
Earned value measurements, 93
Effort hours, 70, 73
Electronic copies of forms, ix
Equations
    beta distribution (three-point estimate), 73,
        76t, 88
    cost estimation, 88, 89t
    duration estimation, 73, 75t
    weighting, 76t, 89t
Estimations
    of cost, 86t, *87,* 89–90t, *91–92*
    of duration, 70–74, 71t, *72,* 75–76t, *77*
    methods, 70, 73, 75t, *77,*85, 88, 89–90t, *91–92*
    of resources, 113
Executing forms, 163–164
    Change log, 175, 176t, *177*
    Change request, 170, 171t, *172–174,*175,
        196, 213
    Decision log, 9, 168, 168t, *169,*196
    Executing process group, 163–164
    Issue log, 9, 165, 166t, *167*

Lessons learned register, 178, 179t, *180,*191,
    229, 230
    Quality audit, 181, 182t, *183*
    Team performance assessment, 184, 185t,
        186–188
Executing process group, 163–164

F
Finish-to-finish (FF) relationships, 67
Finish-to-start (FS) relationships, 67
FLAP approach, 246
Forecasting costs, 207t, *208*

G
Gantt charts, 78, *80*

I
Indexes, performance, 207t, *208*
Initiating forms, 1–16
    Assumption log, 9–10, 10t, *11,*63, 148
    Initiating process group, 1
    Project charter, 2–3, 3–4t, *5–8,*45,
        235, 240
    purpose, 1
    Stakeholder analysis, 12, 15, 15t, *16*
    Stakeholder register, 12–13, 13t, *14,* 15
Initiating process group, 1
Interpersonal competencies, 184, 185t
Inter-requirements traceability matrix, 40, 42t
Issue log, 9, 165, 166t, *167*
Issues, defined, 165
Iterations plans, 45

L
Lag, 67
Lead, 67
Lessons learned register, 178, 179t, *180,*191, 229,
    230
Lessons learned repository, 230
Lessons learned summary, 229–231, 231–232t,
    233–234
Life cycle phase predictions, 28, *29*
Location management, 118, 155
Logs
    Assumption Log, 9–10, 10t, *11,*63, 148
    backlogs for products, 63, 239t, 242,
        242t, *243*
    Change log, 175, 176t, *177*
    Decision log, 9, 168, 168t, *169,*196
    Issue log, 9, 165, 166t, *167*

M

Management Body of Knowledge (PMBOK®
    Guide) – Sixth Edition. *See* PMBOK
    Guide – Sixth Edition
Management reserve, 93
Milestone chart, 79, 81
Milestone list, 65, 65t, 66
Monitoring and control forms, 189–228
    Change request, 170, 171t, *172–174,*175,
        196, 213
    Contract closeout report, 222, 223t, *224–225*
    Contractor status report, 213, 214t, *215–217*
    Earned value analysis, 206–207, 207t, *208*
    Monitoring and controlling process group,
        189–190
    Procurement audit, 218, 219t, *220–221*
    Product acceptance form, 226, 226t, *227*
    Project status report, 121, 191, 196–197, 198t,
        *199–201,*213
    purpose, 189–190
    Risk audit, 209, 210t, *211–212*
    Team member status report, 191–192,
        192t, *193–195,*196
    Variance analysis reports, 202–203, 203t,
        204–205
Monitoring and controlling process group,
    189–190

N

Name, of activities. *See* Activity attributes
Network diagram, 67–68, *69*
Nonlinear numbering structure, 142

O

Opportunity management, 209
Outside contractors, 104

P

Parametric estimates, 70, 73, 75t, *77,*85, 88, 89t,
    91t
Participation, levels of, 101, 102t, *103*
Performance reports
    Earned value analysis, 206–207, 207t, *208*
    Performance indexes, 207t, *208*
    Performance measurement baseline, 20, 83t,
        *84,*93
    project status, 121, 191, 196–197, 198t,
        199–201, 213
    for team members, 191–192, 192t, *193–195,*196
    on team performance, 184, 185t, *186–188*

Planning forms, 17–19
    Activity attributes, 62–63, 63t, *64*
    Activity list, 59–60, 60t, *61*
    Change management plan, 25–26, 27t, *27*
    Communications management plan, 118–119,
        119t, *121*
    Cost baseline, 20, 22t, *24,*93, *94*
    Cost estimates, 85–86, 86t, *87*
    Cost estimating worksheet, 88, 89t, 90t, *91–92*
    Cost management plan, 82–83, 83t, *84*
    Duration estimates, 70–71, 71t, *72*
    Duration estimating worksheet, 73–74,
        75–76t, *77*
    Milestone list, 65, 66t, *66*
    Network diagram, 67–68, *69*
    Planning process group, 17–19
    Probability and impact assessment, 137,
        137–139t, *140–141,*142
    Probability and impact matrix, 142, *143*
    Procurement management plan, 147–148, 149t,
        150–151
    Procurement strategy, 152–154, 153t, *154*
    Product backlog, 35, 59, 63, 239t, 242,
        242t, *243*
    Product vision, 239t, 240, 240t, *241*
    Project management plan, 20–21, 21–22t,
        23–24 (*See also* project management plan)
    Project roadmap, 28–29
    Project schedule, 78–79, *80–81*
    Project scope statement, 3, 45, 46t, *48,*52
    purpose, 17–19
    Quality management plan, 95–96, 96t, *97–98*
    Quality metrics, 99, 99t, *100*
    Requirements documentation, 37–38, 38t, *39*
    Requirements management plan, 33–34, 34t,
        35–36
    Requirements traceability matrix, 40–41, 42t, *43*
    Resource breakdown structure, 116, *117*
    Resource management plan, 82, 104–105, 106t,
        107–108
    Resource requirements, 113–114, 114t, *115*
    Responsibility assignment matrix (RAM),
        101–102, *103*
    Risk data sheet, 144–45, *146*
    Risk management plan, 121–122, 123t,
        *124–127,*137
    Risk register, 128–129, 129t, *130,*142
    Risk report, 131–132, 133t, *133–136*
    Schedule baseline, 20, 22t, *24*
    Schedule management plan, 56, 57t, *58*

Planning forms (*continued*)

Scope baseline, 20, 22t, *24, 49,* 52–53

Scope management plan, 30–31, 31t, *32*

Source selection criteria, 155–156, 156t, *157*

Stakeholder engagement plan, 158–159, 159t, 160–161

Team charter, 109, 110t, *111–112*

WBS dictionary, 52–53, 54t, *55*

Work breakdown structure (WBS), 49–50, 50t, *51,52*

Planning process group, 17–19

PMBOK Guide – Sixth Edition

Close project or phase (process 4.7), 230, 235

Collect requirements (process 5.2), 38, 40

Control procurements (process 12.3), 213, 218

Control risks (process 11.7), 209

Create WBS (process 5.4), 49, 53

data analysis technique, 202, 206

Define activities (process 6.2), 59, 62, 65

Define scope (process 5.3), 45

Determine budget (process 7.3), 93

Develop project charter (process 4.1), 2, 9

Develop project management plan (process 4.2), 20

Develop schedule, 78

Develop team (process 9.4), 184

Direct and manage project work (process 4.3), 165, 170

Estimate activity durations (process 6.4), 70, 74

Estimate activity resources (process 9.2), 113, 116

Estimate costs (process 7.2), 85, 88

Identify risks (process 11.2), 131, 144

Identify stakeholders (process 13.1), 12, 15

Manage project knowledge (process 4.4), 178

Manage quality, 181

Monitor and control project work (process 4.5), 191, 196

overview, viii

Perform integrated change control (process 4.6), 175

Perform qualitative risk analysis (process 11.3), 137, 142

Plan communications management, 122

Plan cost management (process 7.1), 82

Plan procurement management (process 12.1), 147, 152

Plan quality management (process 8.1), 95

Plan resource management (process 9.1), 104, 109

Plan risk management (process 11.1), 121

Plan schedule management (process 6.1), 56

Plan scope management (process 5.1), 30, 33

Plan stakeholder engagement, 158

Sequence activities (process 6.3), 67

Validate scope (process 5.5), 226

Probability and impact assessment, 137, 138–139t, *140–141,*142

Probability and impact matrix, 121, 142, *143*

Procurement audit, 218, 219t, *220–221*

Procurement management

audits, 218, 219t, *220–221*

plans for, 147–148, 149t, *150–151*

Procurement management plan, 147–148, 149t, *150–151*

sources, 155–156, 156t, *157*

strategies, 152–154, 153t, *154*

Procurement strategy, 148, 152–154, 153t, *154*

Product acceptance form, 226, 226t, *227*

Product backlog, 59, 63, 239t, 242, 242, *243*

Productivity rate, 73

Product verification, validation, and acceptance form, 226

Product vision, 239t, 240, 240t, *241*

Progress reports. *See* Status reports

Project charter, 2–3, 3–4t, *5–8,*45, 235, 240

Project documentation

Agile techniques for, 239–248

closing process, 229–232 (*See also* Closing process)

electronic copies, ix

executing forms, 164–188 (*See also* Executing forms)

initiating forms, 1–16 (*See also* Initiating forms)

monitoring and control forms, 189–228 (*See also* Monitoring and control forms)

overview, viii–ix

planning forms, 17–162 (*See also* Planning forms)

Project management plan

component forms, 30, 56, 118–119

elements, 21–22t

form, *23–24*

overview, 20–21

Project roadmap, 28–29

Project performance report, *199–201*
Project or phase closeout, 222, 235, 236t, *237–238. See also* Closing process
Project roadmap, 21, 28–29
Project schedule, 78–79, *80–81*
Project scope statement, 3, 45, 46t, *48,52*
Project status report, 121, 191, 196–197, 198t, *199–201,*213
Project vision, 239t, 240, 240t, *241*

Q
Qualitative risk analysis, 137
Quality management forms
    Procurement audit, 218, 219t, *220–221*
    Quality audit, 181, 182t, *183*
    Quality management plan, 95–96, 96t, *97–98*
    Quality metrics, 99, 99t, *100*
    Variance analysis reports, 202–203, 203t, *204–205*
Quantitative methods, 70, 73, 85, 88

R
RACI chart, 101, *103*
Release plans, 239t, 244, 244t, *245*
Requirements documentation, 37–38, 38t, *39*
Requirements management plan, 33–34, 34t, *35–36*
Requirements traceability matrix, 40–41, 42t, *43*
Resource breakdown structure, 116, *117*
Resource management plan, 82, 104–105, 106t, *107–108*
Resource requirements, 113–114, 114t, *115*
Responsibility assignment matrix (RAM), 101, 102t, *103*
Retrospective, 239t, 246, 246t, *247–248*
Risk assessment, probability and impact, 121, 137, 138–139t, *140–141,*142, *143*
Risk audit, 209, 210t, *211–212*
Risk data sheet, 144, 145t, *146*
Risk management plan, 121–122, 123t, *124–127,*137
Risk register, 128–129, 129t, *130,*142
Risk report, 131–132, 133t, *133–136*
Roadmap, 22, 28–29
Rolling wave planning, 56, 71, 86

S
Schedule management
    baselines, 20, 22t, *24*
    Earned value analysis, 206–207, 207t, *208*
    Network diagrams, 67–68
    plans, 30–31, 31t, *32,* 35
    Project status report, 121, 191, 196–197, 198t, *199–201,*213
    Release plans, 239t, 244, 244t, *245*
    Schedule management plan, 56, 57t, *58*
    Variance analysis reports, 202–203, 203t, *204–205*
Schedule performance index (SPI), 207t, *208*
Schedule tool, 59, 62, 65
Scope management
    analysis, 203, 206–207, 207t, *208*
    baselines, 20, 22t, *24,* 49, 52–53
    plans, 30–31, 31t, *32,* 33
    statements, 3, 45, 46t, *48,*52
    validation, 226, 226t, *227*
Source selection criteria, 155–156, 156t, *157*
Stakeholder analysis, 12, 15, 15t, *16*
Stakeholder engagement plan, 158–159, 159t, *160–161*
Stakeholder register, 12–13, 13t, *14,* 15
Stakeholder relationship map, 158
Starfish approach, 246
Start-to-finish (SF) relationships, 67
Start-to-start (SS) relationships, 67
Statement of work, 3, 52, 53, 152
Status reports
    contractors, 213, 214t, *215–217*
    projects, 121, 191, 196–197, 198t, *199–201,* 213
    team members, 191–192, 192t, *193–195,* 196
Subsidiary management plans, 20–21, *23*

T
Team charter, 109, 110t, *111–112*
Team member status report, 191–192, 192t, *193–195,*196
Team performance assessment, 184, 185t, *186–188*
Team resources, 113, 116
Technical competencies, 184, 185t
Three-point estimate (beta distribution), 70, 73, 75–76t, *77,*85, 88, 89t, *91*
Timelines, 28, *29,*242

Time-phased budget, 93
To complete performance index (TCPI), 207t, *208*

V
Variance analysis reports, 202–203, 203t,
    204–205
Vendors. *See* Contracts and contracting process

W
WBS dictionary, 30, 52–53, 54t, *55*
Weighting equation, 76t, 89t

Work breakdown structure (WBS), 49–50, 50t,
    51,52
Work package, 49, 50t, *51,52*
Work performance reports
    Earned value analysis, 206–207, 207t,
        208
    project status, 121, 191, 196–197, 198t,
        199–201, 213
    for team members, 191–192, 192t, *193–195,*196
    on team performance, 184, 185t,
        186–188